Herrera's 'Plasmogenia' and Other Collected Works

Henderson James Cleaves • Antonio Lazcano
Ismael Ledesma Mateos • Alicia Negrón-Mendoza
Juli Peretó • Ervin Silva

Herrera's 'Plasmogenia' and Other Collected Works

Early Writings on the Experimental Study of the Origin of Life

Springer

Henderson James Cleaves
Earth-Life Science Institute
Tokyo Institute of Technology
Tokyo
Japan

Antonio Lazcano
Facultad de Ciencias
Universidad Nacional Autónoma de México
Ciudad Universitaria
Distrito Federal
Mexico

Ismael Ledesma Mateos
Facultad de Estudios Superiores Iztacala
Universidad Nacional Autónoma de México
Tlanepantla
Estado de México
Mexico

Alicia Negrón-Mendoza
Instituto de Ciencias Nucleares
Universidad Nacional Autónoma de México
Mexico D.F.
Distrito Federal
Mexico

Juli Peretó
Institut Cavanilles de Biodiversitat
i Biologia Evolutiva
Universitat de València
València
Spain

Ervin Silva
Universidad Nacional Autónoma de México
Mexico D.F.
Distrito Federal
Mexico

This work collects a translation of Nociones de Biología, originally published in 1904 by Imprenta de la Secretaria de Fomento, Mexico, a reprint of Plasmogeny, originally published as a chapter in Colloid Chemistry: Theoretical and Applied, Volume 2: Biology and Medicine, in 1928 by The Chemical Catalog Company, Inc., and a translation of La Plasmogenía: Nueva Ciencia del Origen de la Vida, originally published in 1932 in Spain by Cuadernos de Cultura, Valencia.

ISBN 978-1-4939-4096-7 ISBN 978-1-4939-0736-6 (eBook)
DOI 10.1007/978-1-4939-0736-6
Springer New York Heidelberg Dordrecht London

Softcover reprint of the hardcover 1st edition 2014

Printed on acid-free paper

Springer is part of Springer Science+Business Media (www.springer.com)

Foreword to Collected Works of Alfonso Herrera

The book you hold contains scientific work long relegated to obscurity in modern biology after Herrera's death in 1942. Herrera's "plasmogeny" fit into a tradition popular from the late nineteenth century up through the 1930s, sometimes referred to as "cell model experiments." Neurophysiologist Reinhard Beutner, for example, in his 1938 book *Life's Beginnings on the Earth* described Hererra's work and similar cell model experiments as informative guides to understanding the basic properties of the simplest living systems and what role osmotic and other physical forces play in producing growth, division, etc. of structures that imitate such lifelike actions. This research tradition was closely related to colloid chemistry, especially from 1910 until the late 1930s, after which the macromolecular approach (what later became "molecular biology") began to eclipse colloid chemical explanations of living phenomena. Since the late 1970s, in light of "autotroph first" thinking and other recent trends in origin of life research, many modern researchers have taken up Hererra's work again and found in it much that was foresighted. Some experiments by Adolph Smith and Gary Steinman using formaldehyde and ammonium thiocyanate were directly influenced by Hererra's work.[1] Sidney Fox also expressed indebtedness to Hererra's insight about the importance of thiocyanate.[2]Thus, this is an opportune moment for the republication (and first English translation) of some of Hererra's works.

Both scientists and historians of science will find this a valuable resource. Independently of modern recognition of his insights, Hererra's work was prominent among thinkers about the origin of life from the 1910s through the late 1930s. Herrera had a wide international correspondence network to whom he circulated his *Bulletin of Plasmogeny*, ranging from Soviet biochemist Alexander Oparin to psychoanalyst-turned-laboratory scientist Wilhelm Reich. Hererra corresponded with Reich in 1939 about Reich's recent "bion experiments." Reich, like Herrera was much influenced by German physiologist Eduard Pflüger and his theory that

---

[1] Steven J. Dick and James E. Strick, *The Living Universe: NASA and the Development of Astrobiology* (New Brunswick, NJ: Rutgers U. Press, 2004), pp. 71-72.

[2] Sidney Fox, *The Emergence of Life* (New York: Basic Books, 1988); see also Stanley Miller, J. William Schopf and Antonio Lazcano, "Oparin's *Origin of Life*: Sixty Years Later," *J. Molec. Evol.* 44: 351-353 (1997).

cyanogen (CN_2) was a key component of the chemistry that produced the first life on Earth.[3]

As noted in the historical introduction to this volume, Hererra was marginalized in Mexican biology after 1929, and many of those who took an interest in his work were critics of the establishment marginalized to some extent in their own countries (Reich, for example, was a Marxist). This did not prevent Hererra's work from being highly regarded abroad, however. Prominent figures like Jerome Alexander in colloid chemistry gave Hererra considerable notice in his own time, and Hererra's final publication was in the prestigious American journal *Science*. And of course the likes of Stanley Miller, Norman Horowitz and Sidney Fox were prominent figures who took an interest in Hererra's work decades after his death.

What insights about surface tension, rudimentary metabolism, the role of Darwinian evolution in the origin of life, etc. can be gleaned from Hererra's writings? How can those writings help us better understand the history of thought about the origin of life from 1900-1940? Read on, and judge for yourself.

By James E. Strick

Dept. of Earth and Environment
Franklin and Marshall College

[3] James E. Strick, *Wilhelm Reich, Biologist* (Cambridge, MA: Harvard U. Press, forthcoming 2014).

Introductory Notes

Alfonso L. Herrera generated a large number of publications during the course of his career. With the recent uptick in interest in research into the origin of life, we present here three of them representing various stages of the development of his thinking, in chronological order of their publication, with the aim of making his works more widely accessible to the non-Spanish speaking readership of the origin of life and history of science communities. *Nociones de Biología* was originally prepared as an introductory university-level biology text and published in Mexico in 1904. *Plasmogeny* (1928) was originally published in English in the United States as a chapter in volume two of Jerome Alexander's four volume compendium on colloid chemistry (*Colloid Chemistry: Theoretical and Applied, Volume 2: Biology and Medicine*). *La Plasmogenía: Nueva Ciencia del Origen de la Vida*, published in 1932, is a considerably shortened version (45 pp.) of a 1926 volume entitled *Una Nueva Ciencia. La Plasmogenía* (446 pp.). Both were published in Spain.

Two of them (*Nociones de Biología* and *La Plasmogenía*) are presented for the first time in English translation. *Nociones* was translated by HJC and proof-read for accuracy by AL, ILM, ANM and JP. *La Plasmogenía* was translated by Fabiola Barraclough and proof-read by the present authors. The chapter from Alexander's second *Colloid Chemistry* volume is presented in its original format. Translations were kept as close as possible to the original intended meaning, keeping the terminology of the time of writing, but worked so as to improve the flow of the modern English translation. The introduction is based on publications authored by ILM and ANM, as translated by HJC and modified by AL.

The authors would also like to acknowledge the support of funding from the Secció de Ciències Biològiques de l'Institut d'Estudis Catalans (JP), the NASA Astrobiology Institute Post-Doctoral Program, the NASA/NSF Center for Chemical Evolution for support during the preparation of this work, and the Earth-Life Science Institute of the Tokyo Institute of technology (HJC), as well as PAPIIT grant No. IN110712-3 and CONACYT grant No. 168579/11 (ANM). The digitized images from *La Plasmogenía: Nueva Ciencia del Origen de la Vida* were generously provided by Soledat Rubio (Càtedra de Divulgació de la Ciència, University of Valencia).

The authors sincerely hope these materials will shed some deserved light on Herrera's work during this widely overlooked era in research and thinking regarding the origin of life.

Contents

Introduction

Charles Darwin's reluctance to discuss the origin of life surprised some of his disciples and even led to some harsh criticism. The list includes Ernst Haeckel, one of his loyal followers, who wrote a few years after the publication of *The Origin of Species* that "[t]he chief defect of the Darwinian theory is that it throws no light on the origin of the primitive organism –probably a simple cell –from which all the others have descended. When Darwin assumes a special creative act for this first species, he is not consistent, and, I think, not quite sincere…" (Haeckel 1862).

However, Darwin was very much aware of the distinction between the issue of the appearance of living beings, and the understanding of the processes underlying their evolution. He had also realized that at that time it was not feasible to undertake the experimental study of the origin of life. As he noted in the 1861 3rd edition of the *Origin of Species*, "I have now recapitulated the chief facts and considerations which have thoroughly convinced me that species have been modified, during a long course of descent, by the preservation or the natural selection of many successive slight favourable variations. I cannot believe that a false theory would explain, as it seems to me that the theory of natural selection does explain, the several large classes of facts above specified. It is no valid objection that science as yet throws no light on the far higher problem of the essence or origin of life" (cf. Peretó et al. 2009).

And yet, *The Origin of Species* provided an evolutionary framework for analyzing the origin of life, and the possibility that living organisms were the historical outcome of the gradual transformation of lifeless matter became widespread after its publication. This approach would soon merge with experimental observations and the search for a physicochemical basis of life that resulted from descriptions of the gelatinous, water-insoluble substance that was found inside all cells which was termed "protoplasm" by the physician Johann E. Purkinje and the botanist Hugo von Mohl, who like others argued that it was the basic component of life (Lazcano 2009).

These ideas, which were based on Thomas Graham's 1861 proposal that the protoplasm was a colloid formed by a homogenous, proteinaceous substance, were understood by many as implying that living organisms were the evolutionary outcome of the gradual transformation of lifeless gel-like matter into protoplasm. This

H. J. Cleaves et al., *Herrera's 'Plasmogenia' and Other Collected Works*,

DOI 10.1007/978-1-4939-0736-6_1, © Springer Science+Business Media, LLC 2014 possibility was rapidly championed by Thomas Henry Huxley, who in his famous Edinburgh 1868 lecture *On the Physical Basis of Life* famously stated that the basic traits of life could be explained in terms of the chemical and physical properties of the molecules that make up protoplasm (Farley 1977; Strick 2000).

The idea that the first forms of life had been nothing but droplets of protoplasm endowed with the ability to fix atmospheric CO_2 and to use it with water to synthesize organic compounds soon followed. As Haeckel argued in his 1884 *The History of Creation*, bacteria represented the transition between the non-living mineral order exhibited by crystals and the complex biomineral structures exhibited by heliozoa, diatoms and other microbes which he located near the base of his evolutionary trees. For Haeckel the scheme was completed with the demonstration of the metabolic flexibility of euglena and other protists, that can rapidly shift from heterotrophic to photosynthetic metabolism, and that he interpreted as an evolutionary vestige of primordial autotrophy.

The mesmerizing reports of life-like behavior in microscopic droplets of different compositions led many to the conclusion that physicochemical models of protoplasm might provide insights into the origin of life (Lazcano 2009). The ideas of Jerome Alexander, Stéphane Leduc, and Alfonso Luis Herrera epitomize this trend. Like many of his contemporaries, the Mexican A. L. Herrera (1868-1942) was convinced that life could be created in the laboratory, and proposed an autotrophic theory he called plasmogeny. A committed evolutionist, Herrera devoted more than 50 years to experimenting with different kinds of substances, attempting to "illustrate the physicochemical concomitants of life" (see footnote 3 in p. 170, this volume).

Herrera was an energetic character of impressive intellectual versatility who can be considered the principal promoter of Darwinism in Mexico. A self-described mason and atheistic evolutionist, his books are a demonstration of the intellectual prowess of a committed naturalist with a broad perspective of the significance of evolutionary theory. Starting in 1898, his attempts to develop a materialistic explanation of the origin and nature of life led him to use mixtures of water and oil (or gasoline) to understand the shape, size and movement of cell-like structures. He would later refine his ideas and develop his theory of plasmogeny, which attempted to explain the origin of primordial photosynthetic protoplasm.

Herrera's life spanned various important eras in Mexican history. The son of an eminent academic and naturalist, he was born in 1868 and raised in a liberal family. The life of relative privilege that he enjoyed as a child and teenager included an exposure to the most significant medical and scientific advances of his time. Herrera grew up at a time in which the social and political changes that Mexico had experienced led to the recognition that laical and secular values were essential for defining the national education system. During the second half of the 19th century Darwin's books were translated, published and extensively discussed. Thanks to his efforts, the teaching of biology in Mexico started within a clearly defined evolutionary framework that was openly secular and in which the influence of Haeckel is evident (Ledesma 2002).

The many changes that Mexico underwent during Herrera's lifetime forced him to be a contemporary of major political and academic characters and diverse value systems, but he remained loyal to the ideas of Lamarck, Darwin and Haeckel, whose views he zealously promoted. His efforts led to the creation of the Zoological Garden of Mexico City, the Office of Biology Studies (Dirección de Estudios Biológicos, DEB), a short-lived Botanical Garden, a proposal for the creation of a National Aquarium in the Gulf of Mexico and the reorganization of the Museum of Natural History, an institution that he transformed into an urban landmark devoted to outreach activities that played a key role in publicly promoting Darwinian perspectives.

Unfortunately the reorganization of what would become the Universidad Nacional Autónoma de Mexico led in 1929 not only to the disappearance of the DEB but also to his relative exclusion from the nascent institutional environment. In spite of his intense work, he was increasingly isolated from the academic establishment and became a minor player in Mexican national science. This did not stop him from pursuing his work on plasmogeny, which he complemented with the strengthening of his academic exchanges with scientists of various countries. The list included A. I. Oparin, whose heterotrophic theory of the origin of life included coacervates as models of precellular structures Lazcano (2010). Unfortunately the extensive correspondence between Oparin and Herrera has been lost[1], but as Oparin mentioned in 1975, Herrera's letters demonstrated his strong dedication to Darwinian perspectives and his secular approach to the study of the origin of life (Oparin, 1978; Del Rio, 1978).

With the exception of Oparin, however, most of the foreign scientists with whom Herrera had strong academic ties were considered marginal in their respective countries. They had no positions of power, but were in fact critics of the establishment with alternative ideas and proposals. The list includes Leduc and Renaudet in France, Jules Félix in Belgium, Delfino in Argentina, Lecha-Marzo in Spain, among scientists in other nations, and their commitment to Herrera's approach led several of them to create short-lived institutes and societies devoted to the study of plasmogeny in their own countries.

Herrera was a staunch, committed Darwinian who included the work of both Haeckel and Lamarck in his evolutionist perspective. Of equal significance, he included the emergence of life in his scheme. In an attempt to understand the chemical origins of protoplasm, during his lifetime he developed an extensive research program that led to the experimental study of numerous compounds and achieved the formation of microscopic structures which he claimed exhibited important similarities to cells, due to their structure, growth, surprising motility and osmotic properties. As reported in his extensive bibliography (Beltrán 1968), these structures were promptly divided into two major groups: the so-called colpoids, which were produced when olive oil, gasoline and other complex molecules were used, and the sulphobes, which resulted from mixtures of ammonium thiocyanate (NH_4SCN) and formaldehyde (H_2CO) (Herrera 1942).

[1] During a visit to Moscow, one of us (A.L.) actually helped Oparin search in his archives for the correspondence he had exchanged. Unfortunately nothing was found, most likely because Oparin and Herrera started their correspondence during the Second World War and soon after most scientific institutes were moved from Moscow to the eastern part of the USSR for their protection.

Several stages can be distinguished in Herrera's scientific career, including his attempts to produce life-like structures from various combinations of inorganic substances and solutions that continue to fascinate contemporary researchers (Cartwright et al. 2002; Garcia-Ruiz et al. 2002). Based on Pflüger's proposal regarding the role of aldehydes and CN-containing derivatives in biological catalysis (Pflüger 1875), he developed his "sulfocyanic" theory of the origin of life (Ledesma 2002). According to this proposal, the physical structure of cellular plasma was derived from sulfur-containing compounds that formed part of a molecular matrix within which the primordial fixation of CO_2 took place via its reduction to H_2CO.

The allotropes of sulfur, which include complex, leaf-like structures, convinced Herrera that this morphogenetic process was related to the emergence of biological order (Herrera 1942). After many trials, Herrera found that the best starting material for the formation of his sulphobes was NH_4SCN, which he dissolved in formalin (an aqueous solution of H_2CO) and spread in thin layers to evaporate (Herrera 1942). In a number of papers he argued that similar processes could have taken place from cyanides present on the early Earth reacting with volcanic sulfur. As he reported in 1942 (Herrera 1942), the reactions of these precursors gave rise not only to several kinds of cell-like microstructures, but also to starch, two uncharacterized amino acids, globules of red, green and yellow pigments, as well as what he described as a 'proteinoid condensation product' Perezgasga et al (2003).

Herrera passed away in 1942. He died working in the small laboratory he had at home, while observing under the microscope the sulphobes and colpoids he had prepared. By then, developments in cell biology and biochemistry had largely led the scientific establishment to discard the idea that protoplasm was the unique element capable of controlling every aspect of cellular life Florkin (1972). Like the many 19th century attempts to describe living systems based on magnetism, surface tension, radioactivity, and other physical phenomena, the description of protoplasm as the basis of life, and the attempts to explain the origins of primordial autotrophic life through it were soon forgotten (Tirard et al., 2010). In the case of Alfonso L. Herrera what remains is not only the memory of his intellectual passion, but also his intense commitment to the teaching and promotion of evolutionary perspectives and his role in the secular interpretation of biological phenomena.

H. James Cleaves (ELSI, Tokyo)
Antonio Lazcano (UNAM, Mexico)
Ismael Ledesma (UNAM, Mexico)
Alicia Negrón-Mendoza (UNAM, Mexico)
Juli Peretó (Universitat de Valencia, Spain)

References

Beltrán, E. (1968). Alfonso L. Herrera (1868–1968): primera figura de la biología mexicana. *Revista de la Sociedad Mexicana de Historia Natural* XXIX: 38

Cartwright, J.H.E., Garcia-Ruiz, J.M., Novella, M.L. and Otalora, F. (2002). Formation of chemical gardens. *Journal of Colloid and Interface Science* **256**: 351–359

Del Rio, C. (1978) Alfonso L. Herrera: la teoría sulfociánica del origen de la vida y la plasmogenia In A. Lazcano-Araujo y A. Barrera (eds) *El Origen de la Vida: Symposium en Homenaje a A.I. Oparin* (UNAM, México), 61–75

Farley J (1977) *The Spontaneous Generation Controversy: from Descartes to Oparin.* Johns Hopkins University Press, Baltimore

Florkin, M (1972) The dark age of biocolloidology. In M. Florkin and E. Stoltz (eds.), *Comprehensive Biochemistry* **30**: 279–283

Garcia-Ruiz, J.M., Carnerup, A., Christy, A.G., Welham, N.J. & Hyde, S.T. (2002). Morphology: an ambiguous indicator of biogenicity. *Astrobiology* **2**: 335–351

Haeckel, E. (1862) *Die Radiolarien (Rhizopoda Radiaria).* Eine Monographie. Druck und Verlag Von Georg Reimer, Berlin

Haeckel, E. (1884). *The History of Creation, or the Development of the Earth and Its Inhabitants by the Action of Natural Causes.* (Paul Kegan Publ, London), volume 1

Herrera, A. L. (1942) A new theory on the origin and nature of life. *Science* **96**: 14.

Lazcano, A. (2009) Mind the gap! *PloS Biology* **7**:e122

Lazcano, A. (2010) Historical development of origins of life. *In* David W. Deamer and Jack Szostak (eds.) *Cold Spring Harbor Perspectives in Biology: The Origins of Life* (Cold Spring Harbor Press, Cold Spring Harbor), pp. 1–16

Ledesma (2002) *Alfonso L. Herrera: el sabio de Ciprés* (Universidad Nacional Autónoma de México, Mexico)

Oparin, A. I. (1978) in Sobre el origen de las primeras formas de vida. In A. Lazcano-Araujo y A. Barrera (eds) *El Origen de la Vida: Symposium en Homenaje a A.I. Oparin* (UNAM, México), 137-144

Peretó, J., Bada, J. L., and Lazcano, A. (2009) Charles Darwin and the origins of life. *Origins of Life and Evolution of Biospheres* **39**: 395–406

Perezgasga, L., Silva, E., Lazcano, A., and Negrón-Mendoza, A. (2003) Herrera's sulfocyanic theory on the origin of life: a critical reappraisal. *International Journal of Astrobiology* **2**: 1–6

Pflüger, E. (1875) Beitrage zur Lehre von der Respiration. I. Über die Physiologische Verbrennung in den lebendigen Organismen. *Pflugen Arch. Ges. Phisiol.* **10**: 641–644

Strick JE (2000) *Sparks of Life. Darwinism and the Victorian Debates over Spontaneous Generation.* Harvard University Press, Cambridge, MA

Tirard, S., Morange, M. and Lazcano, A. (2010) The definition of life: a brief history of an elusive scientific endeavor. *Astrobiology* **10**: 1003–1009

CONCEPTS IN BIOLOGY

BY ALFONSO L. HERRERA.

Ou va I'homme sur la Terre?
VICTOR HUGO.
MEXICO
PRINTING OFFICE OF THE SECRETARY OF DEVELOPMENT
Calle de San Andres No. 15.
1904
DEDICATED TO PROFESSOR
D. ENRIQUE C. RÉBSAMEN
DIRECTOR GENERAL OF NORMAL EDUCATION
With my deep respect and gratitude.
The Author

INTRODUCTION.

Professor Henry C. Rébsamen, General Director of Normal Education, as proof of his great efforts for the advancement of the youth education, started the class in Concepts in Biology, doing us the honor of proposing a class in this interesting subject, and obtaining from the Ministry of Justice and Public Instruction their gracious respective appointment.

As we do not know of a biology text suitable for our current program, it was necessary to compose the one here published today, almost as it was dictated to the students, the works as listed below having served to prepare our lessons:[1]

"Biology," by Letourneau.
"Darwinism," by Ferriere.
"Concepts of Biology," by Conn.
"Principles of Biology," by Spencer.
"The Origin of the Species," by Darwin.

[1] This little book is in fact a summary of our studies since the year 1888.

H. J. Cleaves et al., *Herrera's 'Plasmogenia' and Other Collected Works*,
DOI 10.1007/978-1-4939-0736-6_2, © Springer Science+Business Media, LLC 2014

"The Descent of Man," by Darwin.
"The variation," by idem.
"Darwinism," by Wallace.
"The Evolution of Sex," by Geddes and Thompson.
"Animal Life," by Jordan and Kellogg.
"The Cell," by Henneguy.
"The Protozoa," by Calkins.
"Botanisches Centralblatt."
"The History of Creation," by Haeckel.
"Anthropogeny," by the same.
"Experimental Science," by Bernard.
"Form and Life," by Houssay.
"The Human Species," by Quatrefages.
"Biochemisches Centralblatt."
Other consulted authorities are mentioned where appropriate.

***

We have taught, with the best of faith, what seems most accepted among the plethora of opinions of leading biologists. We have treated our own personal ideas as tentative and speculative, leaving students the burden of judging them, and rejecting them if necessary.

With regard to the beliefs and philosophical tendencies of the students themselves, we made efforts that they not suffer, because we are amazed by the inaccessible, and for our part we have been reserved to an exaggerated degree, in the treatment of everything not derived from observation and experiment.

It is likely that this first attempt will soon undergo a series of corrections and be followed by another less imperfect edition, which will be published as soon as possible, happily taking the advice and censure of biologists into account.

Before concluding this brief introduction, we would like to express our gratitude to General Don Manuel González Cosío, Secretary of Development, Colonization and Industry, who arranged to have this little book printed in the typographic workshops of their important ministry, and Professor Don Enrique C. Rébsamen, our dear and wise director, who provided the necessary elements for photographic reproduction of the figures, the paper, the most valuable and accredited reference books, and in a word, everything that was thought to be conducive to our proposed goal, that a print run of 200 copies of "Concepts in Biology" be made and that they be dispensed free of charge among the students and national teaching facilities.

Mexico, October 4, 1903.

PLAN OF THE WORK.

BOOK ONE.

FUNDAMENTAL PROPOSITION.

All material phenomena of organisms, in the past and present, have been or are caused by known physico-chemical forces.[2]

Biology is the science of these phenomena.

BOOK TWO.

DEMONSTRATION.

A. Facts of fundamental unity.
B. Facts of cellular life.
C. Facts of evolution.

BOOK THREE.

CONCLUSIONS.

1. What is the future of man on Earth?
2. What is the future of matter in the Universe?

BOOK ONE.

FUNDAMENTAL PROPOSITION.

All material phenomena of the organism in the past and present, have had or are caused by known physical and chemical forces.

Biology is the science that studies these phenomena.

DEFINITION, PURPOSE, AND UTILITY OF BIOLOGY.

The term biology was first coined by Lamarck and Treviranus: it is the important science of life, a beautiful and profound science, which underwent remarkable development in the 19th century, being called upon to explain the origin, form, variety

[2] We refer to material phenomena, without denying or affirming the existence of the intangible.

and structure of beings, colloidal minerals that have inhabited the planet for millions and millions of years, and perhaps, why not admit it, of those which should inhabit distant celestial bodies, and that astronomy together with spectroscopy suggest as a kind of chemistry of the Universe.[3]

Auguste Comte gave the word biology a vast, encyclopedic meaning, as it had to cover an impressive group of sciences, including anthropology.

Letourneau, as noted above, limits it to the explanation and coordination of all the great facts and laws of life, almost like what is commonly understood by general physiology, applying that designation to the two organic kingdoms, "how organized beings nourish themselves, how they grow, reproduce, move, feel and think."

But Pascal has claimed that definitions are an illusion of our spirit and this phrase should be remembered when trying to define the word biology or even set its limits, without first defining the term life.

What is life? No one can say yet. We can provide an entirely physico-chemical explanation of vital phenomena, but if we are honest, it is our duty to caution that until the synthesis of a living being is accomplished, nothing can be certain, no assurance can pervade the scientific world, which is not content with words such as biogen,[4] chromatin or plastin, which in fact explain absolutely nothing.

Our theory that inorganic protoplasm is an osmotic electrolithic device, could resolve this difficult question, but it has still not received the necessary confirmations by competent researchers and thus we can only present it as provisional.

It is thus worthwhile to discuss here the definitions that have been given for life and the place Biology may hold in the updated classification of sciences.

***

Definitions of life.

According to Bernard (Fig. 1) the following principle definitions have been given:[5]

Life is the resistance of organized matter to the causes which constantly tend to destroy it.-*Pelletan.*

Life is a force that resists to the laws governing inert matter.-*Cuvier.*

Life is a combustion.-*Bernard.*

Life is a decomposing material.- *Mitscherlich.*

Life is death, the destruction of tissues by combustion.-*Bernard.*

Life is a minotaur, which devours organisms.-*Buffon.*

Life is a defined combination of heterogeneous changes, both simultaneous and successive.-*Spencer.*

Life is the opposite of death.-*Enyclopédie.*

Life is organization in action.-*Béclard.*

Life is the special activity of organized beings. –*Duges*

It is a principle within action.-*Kant.*

---

[3] The density of the Moon is very similar to that of certain manganese or iron phosphates. Meteorites contain phosphorus and calcium, which are necessary for life [*Guillemin. Le Ciel*, p. 4171). According to Huggins, calcium exists in the Sun. ("*Nature.*" 1898, p. 182.) Apatite, impure calcium phosphate, has been found in the Juvinas meteorite. (*Daubrée. Les eaux souterraines*, t.3, p. 339.) Silicon is not an exclusively terrestrial element.

[4] O. Loew. Zur Theorie der Primären Protoplasma-Energie. "Biologischen Centralblatt." 15 Nov. 1902, p. 736.

[5] Bernard. La science experimentale, pp. 149 to 212.

Fig. 1. Claude Bemard.-French philosopher and physiologist. He was born July 12, 1813. He died February 11, 1878. He demonstrated the unity of the phenomena of life and established the main principles of life's natural physico-chemical explanation.

***

Bernard criticizes all of these definitions and concludes that life cannot be defined and that *its principle characteristics are the creation and destruction of organic matter.*

In our concept a formula, rather than a definition, can be proposed, which is entirely positive:

Life consists of the physico-chemical activity of protoplasm or a specially constituted emulsion with a fundamental condition: osmotic flows.

If these are permanently paralyzed *death* results.

If these are slowed for a short time, *sleep* results.

If these are slowed by cold, *hibernation* results.

If they are slowed by dessication, *dormant life* results.

If they are slowed by excessive heat, *aestivation* results.

In full activity, this is called active or manifest life.

Note that other definitions do not mention the *protoplasm*, the physical basis of life, without which all definitions are inadequate, thus they begin by considering life as an extraordinary phenomenon without primordial basis.

Fig. 2. Active life-Circulation of liquids and gases in the plant. Roots absorb water and dissolved salts, the leaves take CO_2 and O from the atmosphere and O and emit CO_2, O and H_2O. All these phenomena are due to the protoplasm; within each plant or animal cell protoplasm is the only living part.

In effect, the transformations of various organic and inorganic materials that the being contains, and their organization itself, is subordinate to the presence of protoplasm and *without it there can be no life*. That is why the fundamental aspirations of the biologist can be reduced to the study of this curious and unique substance.

In summary:

Life is the activity of protoplasm.

Later we will explain the details of this activity, which seem to be rather simple, but since then, and as a logical conclusion of the above, we can say that:

The object of *biology is the study of protoplasm, in all its manifestations. It could also be called general plasmology.*

These are physico-chemical manifestations, and biology is the science of the phenomena of the organism, which in the past and present have been and are due to known physical and chemical forces, a principle which will be demonstrated in the second book.

***

To give a metaphorical notion of life and manifest instability, it could reasonably be stated:

Life is a prolonged agony.

Indeed, the inevitable aging of tissues, the obliteration of certain channels, the disappearance of fetal characteristics, then infantile ones, then adult ones, are so many more deaths which occur in our body. Everywhere there are dead faculties, white hair, defunct sexual organs.

We are slowly dying. Even in the brain facts and figures are forgotten, and the old person that relates with difficulty his monotonous stories, pausing with every step to remember and even to despair for his lost memory, suggests the idea of an old machine, broken by the wear of gears and shafts.

And what about faculties in general, illusions, activity, love, patriotism! Everything is irretrievably dying.

A true science cannot stop there. *Even a cadaver has its biology*: these phenomena will be explored in more detail later, using more vast and general science, which doesn't make an absolute distinction between life and death, similar to the various chemical reactions due to the action of the atmosphere and water on rocks.

***

The utility of biology.

Thc utility of biology is undeniable.

"Know thyself" is the motto of the truly wise and as yet we only know ourselves in a very imperfect manner.

Medicine is the science of the ignorant, ironically, though it should be the sovereign science, because everything, absolutely everything that man does depends on nutrition, at least from the modern physico-chemical point of view.

Lately, in effect, traits which appear less than material have been associated with the constitution, including those such as character, passions, crimes, temperament. It has been argued that the murderer is sick, atavistic, anemic, degenerate, and that study of hygiene and rational practice, and scientific regulation of marriage in order to prevent harmful unions, will lead to the regeneration of humanity, if only future biologists penetrate deeply enough into the mysteries of life, in order to modify our development.

The applications of biology are also crucial to agriculture. The detailed action of fertilizers remains uncertain, and all cultural practices, which were once so empirical, are being perfected according to modern scientific principles.

"The problem of food and its increasing scarcity will also have a definitive, practical solution, when protoplasm can finally be imitated in such a way that the nutrients are produced on a large scale, using highly economical, natural processes, which are much cheaper than a general chemistry lab."[6]

And without going back to those applications of biology *in the future*, we can be sure that its present utility is considerable.

As we shall later, one of the chapters of biology, the study of fundamental unity, it seems, is the necessary foundation for a new philosophy, as almost all the former are based on details and trifles, without encompassing *all of nature.* It is likely that the new path is coming to discern something of the origin and purpose of things, something like a new kind of light.

---

[6] Berthelot. Les aliments synthétiques. Revue Scientifique, 1896, p. 132.

The naturalist philosopher Haeckel, gradually ascending the scale of his meditations, which began by considering the Monera, very simple organisms, has come to other great ideas and deep as that of monism or a single explanation for all phenomena in the Ether, which gives a kind of total awareness.

For his part, the noted astronomer Newcomb, has dared to publish a dissertation on "the Universe considered as a living organism."

Archibald Geike, Meunier and ourselves, in our modest series of investigations, have also considered the Earth we live on as a body endowed with endless motions and transformations.

Biology is a new science.

Biology is a new science, not because the objects with which it deals are new, but because they are being considered from a new point of view. Animals and plants have long been studied, but not as they are today. The difference between the old and modern methods could perhaps be explained graphically, saying that before organic beings were studied at rest, and now they are studied *in action*. Zoologists and botanists of past ages were limited to looking at plants and animals as specimens for museums,[7] where they were ordered and classified with arbitrary names. The biologist of our times sees the same objects as active, as part of a whole that is always being modified and changing. Researchers of fifty years ago were dedicated to natural history, organic realms were made up of individuals that *had to be classified*: for the biologist of today, individuals *must be explained.* (Conn.)

In summary:

The usefulness of biology is the following:

(1) *Medicine, the deep understanding of the organism.*
(2) *Agriculture, scientific cultivation of fields and the perfecting of farm animals.*
(3) *Sociology, the vices and solutions of society considered as an organism. Educational anthropology. The study of temperament and tendencies.*
(4) *The problem of food and food shortages.*
(5) *The unitary or monistic philosophy.*
(6) *The study of organisms in motion.*
(7) *The explanation of living beings.*

THE PLACE OF SCIENCE IN BIOLOGY

The unity of purpose in all knowledge.

All sciences are instruments focused on the same course, at the same star: *the truth.*

They all fundamentally investigate *quantification*, since everything tangible is reduced, according Stallo, to mass and movement.[8]

[7] See: A. L. Herrera. Les Musées de l'avenir. Mémoires de la Société Alzate, t. IX, p. 221.

[8] In regard what is called matter, there is nothing in the brain that has not been first in the senses, and what operates in these is always a vibration, or a movement (light, sound, shock).

Determining the direction balloons will move is equivalent to studying the amount of force needed to hold them static, or rise in the atmosphere given the volume and weight of such a device. Calculating the orbit of a comet is equivalent to studying the magnitude of their attractions, distance, and speed. Investigating the toxicity of strychnine, per kilo of a poisoned animal, or the toxicity according to molecular weight is also a matter of quantity.

Various instruments and apparatus are used to estimate relative quantities:

Clinical thermometer and other species.

Barometer.

Pneumograph.

Balance.

Esthesiometer (to measure the sensitivity of the skin).

Dynamometer.

Micrometer.

Cathetometer.

Calorimeter.

Gasometer.

Blood cell counter.

Manometer.

Hematoscope.

Biology studies, for its part, a problem of quantity, because consists of chemical reactions behaving according to the laws of the combinations.

It is the mathematics of colloidal bodies, of protoplasmic emulsions.

It is a sub-field of chemistry that carefully considers phosphates, albumins, and the inorganic and organic bodies that form protoplasm.

Going further in this direction, we can ensure that, as it is a sub-field of chemistry, itself a sub-field of general mechanics, as it studies the movements of atoms carried by the eddies of combinations, biology becomes a sub-field of general mechanics.

In the not too distant future, the differences now considered to be essential between the various sciences will disappear, the truly wise will cover their highlights, and chemists, zoologists, physicists, who are blind and exclusively devoted to a single specialty, will no longer dominate, but there will be many generalists dedicated to the fundamental, the study of the linkages of the major hypotheses to experimental demonstration of the principle of universal unity.

The classification of sciences.

All classifications are imperfect, because everything is tied to fundamental unity.

Nevertheless, in practice it is necessary to establish some groupings, in ideas, in objects, in science, bearing in mind that any grouping is necessarily human, artificial and imperfect.

There being in the knowable world only mass and movement, it is absurd to seek absolute separations in its more or less transient modulations or aspects, and say, for example, that a bird is absolutely, completely and definitely different from a meteorite or a light beam; bird, aerolite, light beam, are varieties of movement, modes of the ether. Matter is nothing but condensed force and all forces depend on movements of the ether.

***

40 years ago *botany* was considered as a science very different from *zoology*: today they are both *biology*.

Man wanted a separate kingdom for himself, the human kingdom: since Darwin we are now classified among the primates, the class of mammals, the vertebrates, the Animal Kingdom. Therefore, anthropology, with its entourage of auxiliary sciences, sociology, ethnography, linguistics, archeology, history, becomes a sub-field of zoology.

In short: *Zoology, Botany, and Anthropology, form a single science, Biology.*

But this is not fundamentally different from *anorganology or the science of inorganic bodies*, go ahead say what are the assumptions that may arise about the necessary link between the living and inorganic bodies. Inorganic bodies probably form the basis of those, of organized beings.

Biology is clearly a branch of physicochemical science, because life is reduced to physico-chemical phenomena, but actually chemistry is not only the investigation of certain atomic movements, subject to the laws of physics, and every day the tendency of scholars to restrict the movements that take place in simple combinations to electrical causes is accentuated.

Physics also covers terrestrial and astral matters, chemistry and biology, subfields of general mechanics.

That is to say, we have returned to our starting point: *mass and motion.* (Fig. 3.)

Subdivisions.

Ethereal Science or General Mechanics or Etherology	Mechanics / General Physics.	Mechanics. / General Chemistry.	Mineralogy. Biology or Plasmology.	Zoology. / Botany.

SUMMARY OF THE FIRST BOOK

Fundamental Proposition.

All material phenomena of organisms in the past and present, have had or are caused by known physical and chemical forces.

Biology is the science of these phenomena.

Life consists of the physico-chemical activity of protoplasm, or a specially-constituted emulsion, and its current exo-osmotic and endosmotic conditions.

If they are slowed by fatigue, for a short time .... Sleep.
If these are slowed by cold, for several days or weeks, ... Hibernation.
If these are slowed the drying, ... Dormancy.

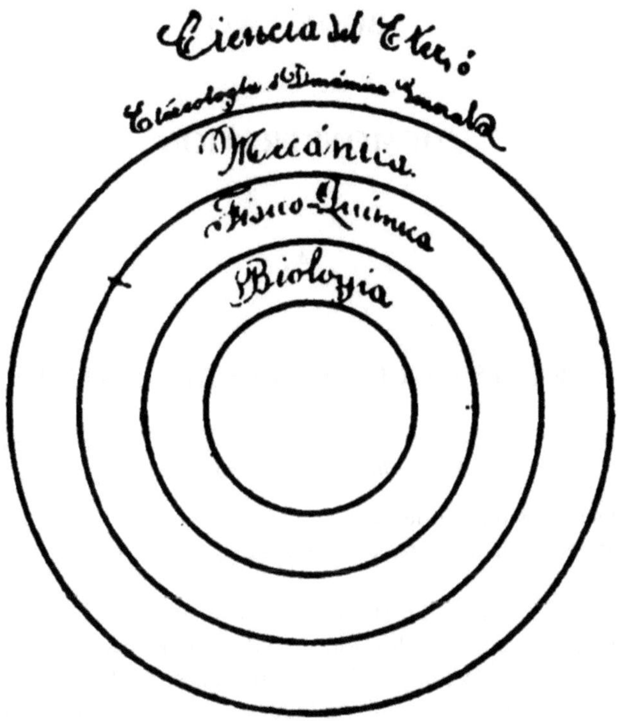

Fig. 3. Subordination and coordination of the sciences.

If these are slowed due to excessive heat,.. Aestivation.[9]
If these are paralyzed by disorganization or other causes,............... Death.

Do these statements apply at once in the same body? Yes. Life is a prolonged agony. There are organisms that sleep. There are dead bodies.

The object of biology: The protoplasm in all its forms and under all possible aspects.

The utility of biology: Very large, for medicine, agriculture, sociology, philosophy.

The place of biology in science:

Subdivisions.

Ethereal Science or General Mechanics or Etherology	Mechanics ——— General Physics.	Mechanics. ——— General Chemistry.	Mineralogy. Biology or Plasmology.	Zoology. Botany.

[9] Probably because excess heat produces thermal tetanus, equalizing interior temperature and slowing osmotic flows.

BOOK TWO

DEMONSTRATION.

A.

FACTS OF FUNDAMENTAL UNITY.[10]

1.-*The importance of this issue.*

D'Alembert stated two centuries ago:

"The Universe, which can be seen in a single glance, is but a single fact and a great truth."

No doubt, adds Geoffroy St. Hilaire,[11] the conception of that illustrious geometer, who was also an eminent philosopher, will never be realized, but science seeks partial demonstration and who would dare to suggest the limit where it will stop? Presently all our knowledge is uniting together with quite numerous and obvious interrelationships, increasingly more narrow and with increasingly logical and necessary ties.

"Whatever their source, they converge towards each other, and therefore advance towards the ultimate truth, as the waters in all parts of the world appear in the surface and gather in the same basin, forming in the beginning countless streams, then majestic rivers, until finally mixing in a single vast sea."

"The philosophical concept of a single truth and therefore of the fundamental unity of science, is not absolutely contradictory to the view that we must put ourselves in the search for truth, and in the study of science. In the *impossible investigation of a single truth, we substitute the difficult investigation of ideas, of partial truths*, which can be regarded as contained in the primary truth. We study, by subordinating according to their hierarchies, in order of generality, from the simplest and most particular to the most complex and most vast; until we arrive at those that in their elevated abstraction, touch the ultimate truth."

A famous writer, a priest whose ideas are not completely separated from Catholic dogma, was forced to conclude, in his famous book "The Unity of the Physical Forces," that "all abstract tendencies, the fine qualities of bodies, the numerous fluids invented to explain physical agents, should be banished from the realm of physics, *because all of the forces of nature depend on movement.*"

---

[10] We also base these demonstrations on some assumptions that seem very plausible as they are supported by well documented facts and not mere doctrinal concerns.

[11] Histoire Naturelle. T. I, pg. 178.

"The objective of science,"says the same scholar," is the knowledge of the physical forces that govern the universe…The reduction of all of the forces of nature to one single principle: that is the tendency of the current scientific trend."

According to Descartes, all variations of matter or the diversity of its forms depend on movement.

According to Kant the resolution of all changes in the material world of movements of atoms caused by constant central forces, would be the complement of natural science.

In short, the great thinkers tend to unify knowledge, forces, matter, bodies, and one of the ideals of the biologist is the association of his efforts with those of so many illustrious philosophers, a necessary and justified aspiration which should not omit a single link in the circular and continuous chain of unity, and if the phenomena of life cannot be reduced to physico-chemical and mechanical laws, there would be reason to doubt human science and all that is under the control of our thoughts. In that case science would be in complete anarchy, like a people who had at once two sovereigns, two codes, two languages, two orders of institutions. But no, all of creation is a republic headed by a democrat, movement.[12]

***

The importance of the fundamental principle of unity is therefore superior to all consideration and facts of unity that can be proved or shown and which are the subject of the following pages, occupy a special place among the tests of our fundamental premise: the phenomena of the organism are due to known physico-chemical forces.

2.- *Fundamental unity of the Macrocosmos, the Mesocosmos and Microcosmos.*

1. Provisionally, to facilitate our studies, we may establish the following groupings:
   The universe or macrocosmos.
   The world or mesocosmos.
   The being or microcosmos.
   Worlds and beings, i.e. the mesocosmos and microcosmos, circulate and live in the macrocosms.
   The microcosmos lives in the mesocosmos, or worlds.
   Atoms and structural and nutrient molecules circulate in the microcosmos. (Fig. 4)
2. Perhaps the Macrocosmos circulates in an even larger and more ethereal medium. But nothing can be said about this presently. The light of some stars that lie beyond the Milky Way takes about 700,000 years to reach Earth. How will man ever pass those limits?
3. The grouping we have made allows several subdivisions:
   Microcosmos of the first degree. – atoms.

[12] It is probable that even the nomenclature of science will be amended in accordance with this trend and we will come to speak of Biomechanics, Physicomechanics, Chemomechanics, Astromechanics, Cosmomechanics, Astromechanics, Geomechanics, and its divisions, Zoolomechanics, Electromechanics, Ethnomechanics, Cytomechanics, etc.

Fig. 4. Elements of the Universe. The question marks correspond to the unknown, which is inaccessible to astronomical research.

Microcosmos of the second degree. – molecules.
Microcosmos of the third degree. – Minerals. Beings.
Mesocosmos of the first degree. – Planets.
Mesocosmos of the second degree. – Solar systems.
Mesocosmos of the third degree. – Nebulae.
Macrocosmos of the first degree. – The Universe.

4. Microcosmos can be called colloidal minerals, hydrosols, plasmas or beings. The same mineral may be crystalloid if it is prepared in the dry state (calcium phosphate) or colloidal (if it is prepared in water). There is no radical difference between the microcosmos, mesocosmos and macrocosmos. There is only mass and motion, expressed in various manners.[13]
5. This assertion is demonstrated very well by the study of Physical Chemistry and comparative Geology. M. Daubrée has been able to imitate many cosmic phenomena in the laboratory, and has established clear principles of the unity of heavenly bodies, especially the meteorites.[14]

---

[13] Huxley states that the microcosmos is an imitation of the macrocosmos and a chain of union and causation links the nebula, and the origin of the sun and the solar system, with the protoplasmic basis of life and the organism. Revue Scientifique (3) T. XXVIII, p. 233.

[14] Daubrée. Etudes synthétiques de géologie expérimentale. A thick tome in 8 °C

6. It could be said that the detectable universe consists of more or less agglomerated microcosms, forming colonies of varying degree of compactness.
7. The fundamental unity of the objects with which we are dealing is demonstrated by various evidences.

1.- *The unity of the physical forces,*

accepted by Secchi and all physicists.

It is clear that since the forces are united this cannot be any other way.

2.- *The conservation of energy.*

It would be incomprehensible in fundamentally different mediums.

3.- *The unity of the forces. Movement.*

"In the world there is only matter and motion, or rather, as the movement is nothing but an act of matter, there is only matter in motion."

"When you say, for example, that heat is converted into electricity, it is understood that the atomic vibration called calorific, changes form and the vibration becomes electric. Weight is changed into heat when a falling body stops suddenly. One could therefore say movement rather than force."

All atoms are in constant movement, those of a metal and those of an organism: we are animated torrents. Sometimes the vibration is obvious due to its great effects and at first sight it is not apparent that there is a corresponding loss of material. For example, a fragment of musk, deposited on a very sensitive balance, produces its special odor without an apparent change in weight.

It has been estimated that the loss of matter from each square centimeter of the surface of radioactive substances, such as compounds of uranium, would be a milligram every billion years.[15]

The English physicist Joule has shown that every body, upon suffering a mechanical force, friction, shock, etc., is heated proportionally to the amount of work done, that is the totality of the movement that has shaken the mass, has been transformed completely in molecular vibration and heat. Therefore the mechanical equivalent of heat can be accurately determined, i.e. how many kilogram-meters are needed to raise a kilo of water by one degree of temperature. A kilogram-meter represents the work required to raise one kilo to one meter in height. Joule found that 424 kg·m were needed

to obtain this result. The mechanical equivalent of heat associated with water is, according to this calculation, 424. Reciprocally one kilogram of water falling freely from a height of 424 meters and stopped abruptly in the fall, will be heated by one degree.

The amount of heat needed to raise the temperature of a kilo by one degree is called a calorie. A calorie is therefore equivalent to 424 kg·m.

It is easy to show, through numerous experiments, that the calorific movement can be transformed into electric, light or chemical energy, or even sound, as evidenced by the wonderful experience of singing flames or har-

[15] Jones. Elements of Pysical-Chemistry, p. 437.

Fig. 5. Father Angelo Secchi, an astronomer and physicist. He demonstrated the unity of the forces.

monic chemistry. Conversely, all of these movements can be changed into heat or into each other. There is thus reason to admit that there are mechanical equivalents of electricity, light, sound, and molecular motions occurring in chemical reactions.

And as is known, the mechanical equivalent of heat is sufficient to transform these various movements in a determined amount, to obtain the equivalent mechanical work. This work has not yet done with sufficient accuracy. Nevertheless, Secchi (Fig. 5) has found a way to approximate the amount of electricity needed to decompose 106 milligrams of water can raise the temperature of 38 grams of the same liquid by one degree.

As we will see thermo-chemistry can be considered the basis of biology.

***

Finally, the principle of conservation of energy is similar to the preservation of life, described by Preyer, and he says in summary: living bodies are in an almost constant relation with a mass of dead matter.

4.- *The unity of matter. The Protile. Corpuscles. The electric atom or electron.*

Wendt believed that all bodies are derived from six key elements and these are derived from a single one.

Crookes had already given, in his "Genesis of the Elements," the name of *Protile*, to this hypothetical primitive matter.

According to Thomson, "most chemists accept that atoms of different elements are different groupings of atoms of the same species." Prout supposed that atoms of different elements were hydrogen, which cannot be accepted, but if instead of hydrogen we say some substance X, nothing could oppose this hypothesis.

"This matter exists in a new state in cathode rays, in which it is much more divided than in the gaseous state, a state in which all matter, that derived from various sources such as hydrogen, oxygen, etc., is of the same class, this matter being the substance of which all of the chemical elements are built."

Thomson called the tiny particles or primordial atoms, into which gas molecules are dissociate, *corpuscles*. Obviously they must be much smaller than the molecules of these gases.

So, the ideal of the alchemists, transforming base metals into gold (the philosopher's stone) has a basis in indisputable truth, and one only needs to know the mechanisms of the deep transmutation of matter.

As it appears, in all probability, that matter and force are a single phenomenon, the assumption that living beings are completely and fundamentally different from their surroundings, from which they form, are fed and sustained, seems absurd. If this were so one would not understand what kind of relationships can exist between such dissimilar things and how to subtract one from the other unavoidable action.

Chemical analysis has shown that all beings are composed of carbon, hydrogen, oxygen, nitrogen and salts.

***

According to Crookes, what we call matter is the effect on our senses of the movement of molecules. At absolute zero temperature, all intermolecular movement disappears and matter as we know it ceases to exist.[16] Each *electron* carries a small electrical current, *if that is not what an electron is. Electrons*, smaller than atoms, can pass through a screen (*Radium* rays) and form the atoms of the various bodies. The *electron* is the only thing that exists in the universe, it is the atom of electricity, the words positive and negative mean there is an excess or deficit of electrons. The electron is equivalent to the Protile. Electricity could be a kind of soul of the universe.

5.- *The unity of organisms. Protoplasm.*

Having established that the cell wall was inactive to an extent, attention became fixed on the contents of the cell. Twenty years after the formulation of cell theory, the nucleus and the cellular substance are considered as essential to the activities of the cell. These ideas appeared indecisive in the writings of several authors, until in 1860 a theory was formulated which has often been called the birth of modern biology. Since then the idea of protoplasm has been raised to great heights. It was

[16] C. R. Acad. Sci Paris. 12 Juillan. 1880.-Modern Views on Matter. The Realization of a Dream. "Science." June 26, 1903.

perfectly defined by Schultze, who argued that the truly active part of the cell was the cellular substance, which was identical in all cells, animal and plant, higher or lower. *It is really the only thing that life consists of.* Despite its simplicity, it possesses all the properties of living beings—irritability, contractability, assimilation and reproduction.—*Huxley called it the physical basis of life*. This concept of protoplasm greatly simplified problems related to the study of life. To understand the nature of this material it was not necessary to study the confusing mass of complex organs, which organic beings possess, nor even the less confusing structures of individual cells.

It is sufficient to study the protoplasm.

Everything in the world of organisms is or has been protoplasm.

For example, the table on which we write is made of wood, whose fibers or cells were full of protoplasm; paper is made of cellulose secreted by the protoplasm; ink, the tannin that it produces in the gills and the bark; the hand, fingers, arm, nerves and vessels, even the brain itself, have no foundation or other fundamental basis.

The protoplasm is the Infusorium, the man, the palm tree, the seed, the insect, everything that lives or has lived. It can be said, as organic chemistry is carbon chemistry, the science of living things is general plasmology or the science of the protoplasm. We have said nothing else since we began these lessons. *Life consists in the activity of protoplasm. Biology is the study of protoplasm in all its forms and under all possible conditions.*

***

Human vanity is built on unsafe ground. We are a type of salts like those that the chemist produces in his glassware and chemical assays, of which a mere whiff is sufficient to reduce us to the immobility of a cadaver. Our protoplasm is extremely delicate: it is composed of solids and water, forming an emulsion, a type of foam, as perishable as the foam of the waves.

Continuity of protoplasm.

Recent studies by microscopists have shown that the walls of cells almost always have small pores, from which exit protoplasmic filaments, which provide an intimate communication between the various parts of the body, *so that the living being is not a plurality, but rather an anatomical and functional unit.* (Fig. 6).

6.-*Parallel between the protoplasm and the Cosmos.*

IN PLURIBUS UNUM.

The Cosmos, just like the protoplasm, is composed of more or less condensed clusters, consisting of elementary parts.

Protoplasm is a hydrosol or emulsion formed according to the theory of Mr. Hardy, by solid particles suspended in a liquid.

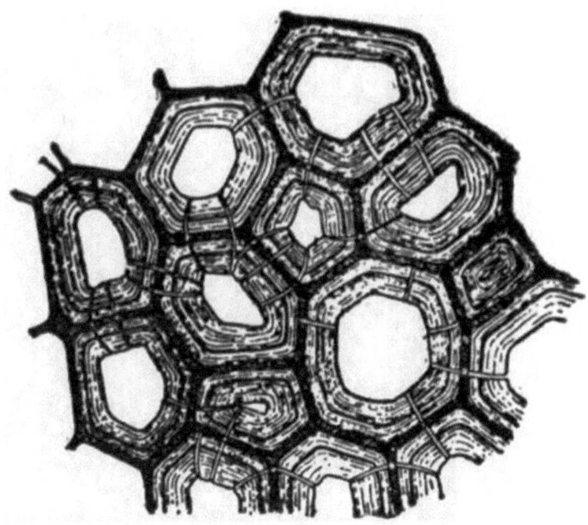

Fig. 6. Communications of the cells through which pass filaments that establish the continuity and unity of protoplasm in the body. (*After Sedgwick and Wilson: General Biology*, p. 12).

Each system of the Cosmos is also composed of cosmic matter or elements (stars, planets, different materials) placed at different distances, more or less compacted together by attractive forces, or rather, with some precipitating in the empty gaps which produce larger ones on their periphery.

The alveoli of an emulsion are united by similar causes (according to Hardy, by electrical causes.)

The double nebula described by Ross shows two central groups with extensions very similar to those of a rhizopods or a multipolar cell. The planetary nebula of Ursa Major has the appearance of two cells set in another larger one with their respective nuclei.

The nebular star in Gemini, like all of its kind, has a bright interior point, like a nucleus, and radiating stripes, similar to the rhizopods of a *Gromia*. The nebular star in the constellation *Argo* also has two nuclei and radiations similar to those of a protozoan, *Cercomonas crassicauda*.

All these celestial bodies are in constant activity, some, like the moon, appear to be dead, but they fall into one another, merging with them and parting again.

According to the hypotheses of modern astronomers, the viscous mass that separates the planets, had an interior movement, which gradually became faster and faster, increasing the temperature differences between the superficial and deep parts, which was finally communicated to the total mass.

Plateau's experiments have made it clear how a planetary system may form, by rotation and breakdown of viscous drops.

Later we will show that our planet and all its internal and external parts are site of currents of very diverse nature, of changes of forces, of incessant transformations, like living protoplasm.

But this is not all: even the phenomena of cell division, of separation by excess nutrition, have their analogies in the world of stars. For example, comets, under very ill-defined conditions, fragment, giving rise to similar bodies.

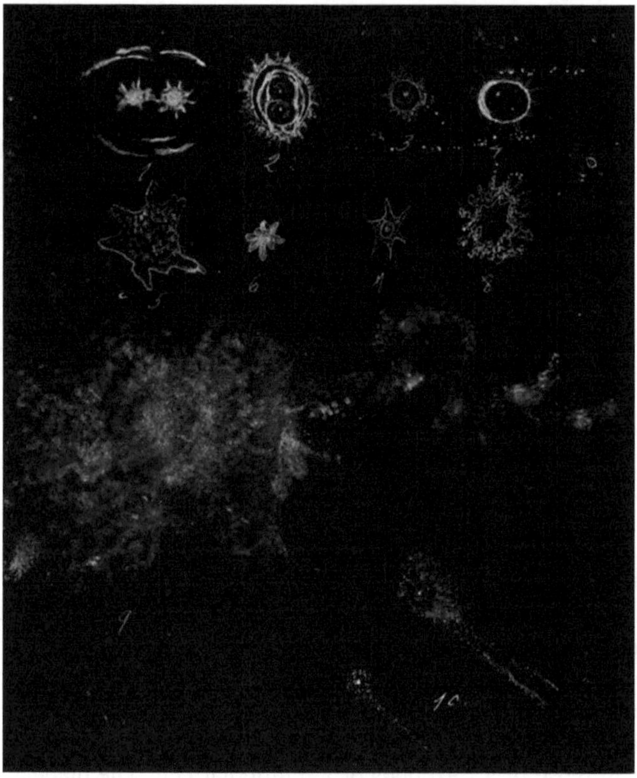

Fig. 7. 1. Double nebula, according to Ross. 2. Planetary Nebula of the Ursa Major. (Roese). 3. Nebular star in Gemini. 4. Nebula Star in Argo. 6. Primitive proto-amoeba. 6. Amoeba diffluens. 7. Bone cell. 8. *Cercomona crassicauda*. 9. Nebula in Argo. 10. Division of Gambart's comet.

Gambart's comet was discovered in 1832. In 1846 it appeared divided into two comets of different sizes, which grew further and further apart. They reappeared in 1852, traveling together, at a distance of 500.000 miles. (Fig. 7).

As the Cosmos is fundamentally unitary and its formation obeys general principles, according to a determined plan, it is logical to expect that its various parts offer analogies, since evidently the same forces, the same causes, hold together the elements of every sphere, be it a world, a snowball or a sea urchin.

Ball lightning, according Préaubert, is very similar to the amoebae: having ameboid extensions which seem to fly and go through obstacles, penetrating through crevices, in a way similar to leukocytes through the walls of a capillary.

It is clear that the Gemini Nebula shows many differences with cells, for example its size, composition, and temperature, but these differences are not absolute and those who understand it this way may never acquire a philosophical idea of things. But there's still more: silicon, the most abundant element on the planet after oxygen, forms extraordinary organic-like structures and is a major contributor to the constitution of planets. Is the protoplasm of the mineral kingdom and some still believe it to be the inorganic basis of living protolplasm (see the facts of cellular life.)

Protoplasm.	*Cosmos. (Mesocosmos).*
Solid and liquid parts of different size. Gases.	Complex composition.
Movements. Movement of forces and matter.	Id.
Delay of movement in certain parts, or in certain specific circumstances.	Id.
Constant transformation of forces, as in any machine.	Id.
Phenomena of division.	Id.
Complex composition.	Id.
Phenomena of condensation. (Nucleus).	Id.
Phenomena of evolution.[a]	Id.
Material existence.	Id.
Material properties.	Id.

[a]According to Crookes, the atoms of chemical elements are probably not eternal and may decline and die: under the influence of electricity, chemical reactions and temperature within the Protyle, where all matter is in the pre atomic-state, forming first the low atomic weight elements, and finally very high atomic weight elements, such as thorium and uranium.

***

Dr. Alfredo Duges has sent us this suggestive news, "When speaking of the correlation between the phenomena of life and the physical world, Agassiz quotes the following comparison of the phylotaxic numbers (order of leaves and cycles of leaves on the stem) with astronomical numbers:

Peirce discovered these harmonies:

Names.	Actual revolution time in days.	Successive time of revolution.	Series of phylotaxic fractions.
Neptune.	60.129	62.000	—
Uranus.	30,687	31,000	½
Saturn.	10,759	10,333	1/3
Jupiter.	4,333	4,133	2/5
Asteroids	1.200 to 2.000	1,550	3/8
Mars.	687	596	5/13
Earth.	365	366	8/13 } 8/21
Venus.	227	225	13/21 } 8/21
Mercury.	88	87	13/34

That is:

31,000: 62,000:: 1:2.
10,333: 31,000:: 1:3.

The movements of celestial bodies and the atoms that combine in the reactions are not essentially different.

Secchi and almost all physicists and astronomers agree that the ether is a resistive medium and its components are falling into each other, with the largest masses producing around them a relative vacuum in which the smaller ones precipitate. The moon is falling towards the Earth with a determined speed.

All the planets are falling into the Sun and this path to their downfall is inevitable and fatal.

Fig. 8. The union of atoms according to the hypothesis of mechanical fields. In the figure above are two atoms, one oxygen and one nitrogen, which join to form a molecule of nitrous oxide, the other two figures show the formation of a water molecule by two hydrogen atoms that fall into one of oxygen.

This is what happens on the surface of water with floating wooden spheres of very different volume, the largest ones produce a depression in the liquid, in which the smaller ones precipitate, as if they were pushed by a mysterious force akin to gravity. If you blow into the mouthpiece of a pipe, which has a small hole in the side, there is a depressurization and the outside air applies a force over the hole, holding up a wooden ball. It seems to be attracted by an unexplained force.

It can be assumed that atoms are constantly vibrating and leave gaps around themselves relative to other atoms that precipitate. This explains the action of heat on combinations, as the vibrations of atoms grow under the influence of heat, they join or become separated (Fig. 8.)

To form a molecule, two atomic masses of two simple bodies require that the contact between them is more or less intimate, which is achieved either through pressure, an increase in velocity by heating, electricity or light, shock, the presence of a porous body such as coal, platinum, manganese, oxidizing ferments, the iron in hemoglobin or the alveoli of protoplasm, and of course the aqueous transport of ions.

Atoms are connected by a variable energy. Upon putting several molecules into contact a crash occurs, a temporary disturbance of equilibrium, and during this new more stable groups are formed. For example, barium nitrate and sodium sulfate, four principle molecules are put into movement upon contact with water (ions) and a double decomposition takes place (sodium nitrate and barium sulfate).

Dolbear proposed in 1889 a similar theory.[17] It supposes that in the ether there are animated annular atoms with rapid alternating movements, expanding and contracting,[18] that determine around each atom a zone of rarefied ether, in which the rarefaction will decrease with distance from the ring.

If that second ring is close to first and enters the rarefied area,it will be pushed by the side where the ether has thinned less and be thrown on the first ring, as if attracted by it. The ring is then surrounded by a zone of attraction or mechanical field. The molecules have a more complicated mechanical field.

According to Dolbear's theory or ours (which are really equal) any new group assumes an increase in mass and a decrease of speed, and thus the same peripheral vacuum and attraction. The combinations are generally defined.

This explains other facts, for example:

The formation of unions by a formidable compression.

The saturation of molecules.

Autosaturation (Carbon.)

Avogadro's law.

Combinations of fusion, dissolution, pressure.

Dissociation followed by recombination in the cold part of an apparatus.

***

Later we will see how to apply these theories to the protoplasmic function.

7.-The thermochemical universal.

The principles popularized by Berthelot show that any combination is theoretically possible when bodies are given the amount of caloric necessary.[19] That is, this leads to the formation of voids or mechanical fields of a value x.

Decompositions require the same condition, but it is necessary that the atoms acquire the exact necessary speed directly, to escape the wells in which they were trapped.

It often happens in practice that secondary reactions take place or the optimum amount of heat is not provided.

However, if the previous concepts about unity are well founded, thermochemistry should be generally applicable to inert matter and living matter. And indeed, according to Berthelot, all living beings exist without consuming an energy that is adequate and the heat produced is equal to the formation of their elements and tissues.[20]

Thus, based on the heats of formation of inorganic or organic compounds, it has been possible to reproduce them by synthesis (for example, neurin, a base which forms part of lecithin and exists in the brain and sperm).

---

[17] Delage. La structure du protoplasm, p. 452.

[18] Corresponding to the atoms composed of electrons of Crookes, and these are probably electric currents.

[19] La synthèse et la termochimie. Revue Scientifique, 25 Novembre 1882, p. 674.

[20] Excluding structures like the protoplasmic reticulum, which should not ever be burned.

In addition, thermal conditions are at the root of all the biological ones and one can say that living beings and their lives are essentially calorific phenomena.

Electricity, electrolysis, and the dissociation of ions have an influence, but at a certain temperature, *it is very probable that some of the heat energy is transformed into electricity*.

The thermal intensity influences plant growth, chlorophyll production, the decomposition of CO_2, the absorption of O, H_2O absorption by roots, etc. (See notes II and III.) The development of silk moth eggs requires not only an optimum temperature, but also a period of minimum temperature. The evolution of locust eggs is proportional to the temperature: 65 days to 10 °C and 60 days at 15 °C.

Ferments. Their action should be primarily thermal. If the transformation of glucose into alcohol and CO_2 requires 71 calories,[21] according to Berthelot, it is clear that the direct or indirect decomposition of alcoholprovides, perhaps by determining successive reactions that give off that quantity of heat, the influence of some mineral element such as the manganese in oxidases, and which, divided, pulverized between organic matter, presents more surface to the action of the agents.

It is likely that alcohol can be made by heating glucose, since the transformation of starch into dextrin can be achieved through a fermentation, amylase, or by roasting with sulfuric acid or potash.

The coagulation of milk is determined by means of enzymes, acids or heat (donkey milk.)

Enzymes can be compared, by the enormous power they develop, to an explosive such as nitro-glycerin, which upon exploding produces 10,384 volumes of gas.

As is known, a platinum sponge has the property of condensing gases, sometimes igniting them (pyrophor.) However, alcohol can be turned into vinegar, by oxidation. H. Nilson has achieved the synthesis and hydrolysis of ethyl butyrate using a platinum sponge.[22]

Finally, inorganic colloidal ferments have recently been obtained, which act like organics, at certain temperatures. (See: predominant role of inorganic substances in biological phenomena.)

Thermal classification of organisms.

Because biology is closely linked to thermochemistry, a new thermal classification of organisms can be established (note V). The general names of eugenesics, maxitherms, etc., would undoubtedly be of an almost mathematical exactitude.

Special action of chlorophyll.

It works like a simple screen and absorbs less refractible and warmer radiationand provides the underlying protoplasm the quantity of heat necessary for the decomposition of CO_2, provided that the radiation is optimal and of short wavelength. It has been replaced with solutionsdiaterman of iodine in carbon disulfide. But it is difficult to provide the optimal temperature. (The decomposition of CO_2 is due to a ferment, according to some authors).

---

[21] We stress that heat can be transformed into electricity. (Thermo-electric battery). Secchi. L'unité des forces physiques, p. 351.

[22] Science. XV, p. 715.

Toxicology.

Rabuteau, since 1867, found that most metals are more toxic the higher their atomic weight, that is, the weaker their specific heat, since the atomic weights of simple bodies are in inverse proportion to their specific heats. Thus, osmium is highly toxic (specific heat = 0.03063) and has an atomic weight of 167, while sodium, so innocent, is very light (atomic weight = 23, specific heat = 0.2934) Note VII.

A curious coincidence: the formation of vaccines and antitoxin sera are usually obtained by cultivation at a certain temperature. The toxins of diphtheria are greatly attenuated at 60 °C or 70 °C, the globulins of snake venoms become anti-toxins at 60 °C. Finally, the morphology of certain bacteria varies with temperature (Bacillus of Roeser. Note VIII.)

***

Artificial protoplasm, regardless of its composition, reacts strongly under the influence of heat, and it shows increased currents and movements in these emulsions, especially in oleates and inorganic silicates. Most likely is that the heat influences and thus mobilizes molecules in natural or artificial protoplasm.

***

HIGHLIGHTS OF THE FACTS OF BIOLOGY RELATED TO THERMOCHEMISTRY.

Wintering. Dormancy.

Nutrient flows are slowed, as we have indicated, and internal condensation of heat stops. During sleep the same happens. All movements of protoplasm require the presence of oxygen, or the heat of combustion.

Death, the final slowing of the currents, coincides with a cold that only putrefying bacteria can resist.

The parthenogenetic generations of *Pediculoides ventricosus*, Acariano, an insect parasite, are suspended when this mite is cultivated in the winter in Mexico.

Aestivation or the lethargic sleep of the Tenrec during intense heat, is perhaps due to a thermal tetanus. This is likely due to thermal uniformity. When the viscous interior and exterior liquids of a cell or any other osmotic apparatus become equal, osmotic flows are suspended.

Nutrition.

Undoubtedly, according to research by W. Roux and others,[23] every biological phenomenon is related to a mechanism of nutrition, from fertilization to ontogenesis and phylogenesis. However it was necessary to establish the physical-chemical

[23] Herrera. L'origine des individus. Memorias de la Sociedad "Alzate." T. XI, p. 137.

principle concerned, which can only be heat.[24] Nutrition,the multiple internal and external relations, intrinsic and extrinsic, between the being and the environment, is *the accumulation of reserves that produce heat.* An organism that ingests 1 g of albumin accumulates 4.998 calories; another that ingests beef fat, 9.069. Plants spend less than animals and produce more under the influence of sunlight filtered through chlorophyll. (Note IX.)

Improvement.

To grab these heat condensors, fats and proteins, the living being needs the senses, arms, claws, nerves, organs and systems, and the more divided they are, the more perfect.

Decline is almost impossible, since the decrease of organs, senses, and faculties cannot provide much nutrition (heat), as the increase and perfection of the organs, senses and faculties.

Human efforts mainly tend to hoarding calories and protection against the cold, which comes in the form of poverty, nakedness, hunger, weakness and disease or calorigenic depletion by excessive work.

The basis of all human industry is fuel, and in the same way, the basis of all protoplasmic industries is fuel and heat, the ultimate form of force.[25]

***

The life and the thermodynamics of the atom, of the living being, of the star, are special cases of the life and thermodynamics of the ether, the cosmic medium which comes from the unknown…

Nevertheless, the modern and fertile ideas about the ionization of molecules tend to give prominence to the production of physico-chemical phenomena by direct or indirect, mediate or immediate electric forces and actions. The true philosopher is always ready to believe these transformations of energy and the true naturalist expects and provides the constant transformation of species.

The real difficulty in the natural sciences is, to remain impartial, not giving predominance to that which is only secondary. In this manner, J. Loeb attributes the action of salts on protoplasm to the electrical charges of the ions, but it is known that electricity is constantly accompanied by thermal work,[26] and it could be that it works more by heat loss and a correlating change in the viscosities of liquids found in the osmotic apparatus, than by a simple electric perturbation.

In summary, these subtleties detract us from the path of generalization that we have identified. It is not possible that the forces which seem so isolated and independent in nature as they are in our minds. We should reduce them to a single expression: movement.

---

[24] Or if you prefer electric, we repeat that in the apparatus-organism forces are transformed, as in an artificial machine.

[25] See Chapter VI of the work of Jones: Elements of Physical Chemistry, p. 279.-M. Berthelot. La chaleur et la masse chimique. Revue Scientifique. 10 Juillet 1880, p. 26.

[26] Secchi, 1. c, p. 287.

NOTE I.[27]

Fundamental principles of thermo-chemistry.

1. All chemical phenomena are phenomenon accompanied by heat.
2. The amount of heat given off by any reaction measures the total amount of chemical and physical work (and could even say vital) verified in that reaction.
3 The amount of heat released depends only on the initial and final state of the chemical system, regardless of the nature and sequence of intermediate states.
4 All verified chemical change without the intervention of an external energy (heat, electricity, light), tends to produce the body or system of bodies that give off more heat.

NOTE II.

Plant growth.

	Optimum
Peas and Lupinus	26 °C6.
Berro and Linseed	27 °C4
Corn	33 °C5
Melon	37 °C2.

Chlorophyll production.

	Minimum.	Optimum.	Maximum.
Barley	4–5 °C	30 °C	37–38 °C
Radish	10 °C	35 °C	45 °C
Corn	10 °C	35 °C	40 °C

Decomposition of CO_2.

Larch	0.5 °C to 2.5 °C
Herbs, grasses	1.5 °C to 3.5 °C
Vallisneria	6 °C
Potamogeton	10 °C to 15 °C

Oxygen uptake.

	Minimum.	Maximum.
Wheat	0 °C	40 °C

[27] These notes are not mandatory for students.

Water absorption by roots.

	Minimum.	Maximum.
Tobacco, pumpkin	3 °C -5 °C	12 °C -18 °C
Nabo	0 °C	
Ivy	0 °C -1	
Rose-bay		16 °C 5

Emission of CO_2.

Wheat	20 °C to 44 °C

Transpiration.

	Minimum.	Maximum.
Valley	20 °C	44 °C

Movement of the protoplasm.

	Minimum.		Maximum.
Ciliary. *Chlamydococcus*	5 °C		43 °C
Interior. *Nitella*	0 °C 5		37 °C
		(Van Tieghem.)	
Tabs of the cilia.	25–30 °C		35 °C

Thermotropism.

Paramecium	0.01 °C and even 0.005 °C

These determine an orientation.

NOTE III.

Development.

	Temperature.	Development.
Frog eggs	15 °C	10 days
	10 °C	21 days
	5 °C	0 days

	Temperature	Days
Frog Metamorphosis:	15 °C	73
	10 °C	235
Crusapus eggs	16 °C to 20 °C	8
	30 °C	1

	Temperature	Days
Lobster eggs	10 °C	65
	15 °C	60
	20 °C	55
	25 °C	50
Life of a copepod	18 °C	77
	26 °C	38[a]

[a] The Physiological Zero and the Index of Development for the Egg of the Domestic Fowl. A Contribution to the Subject of the Influence of Temperature on Growth, by Ch. L. Edwards. American Journal of Physiology. Vol VI, no. VI, 1902.

Hibernation.

Batracians, Reptiles	5 °
Molhelis pomatia (I. G. P.)	5 °
Insects and caterpillars......	3 ° to 4 °
	(Cuénot.)

NOTE IV.

Principles relating to living things according to Berthelot.

The heat generated by a living being in any period of its existence, without the aid of a external energy, other than their food (+O and H_2O) is equal to the value produced by the chemical metamorphosis of immediate principles in their tissues and food, minus the heat absorbed by the external work performed by the same being.[28]

Life is sustained and does not consume any energy to it their own.

NOTE V.

Thermal classification of some microbes.

Maxitherms.	41 °C to 60 °C
Bacillus subtilis, tuberculosus, anthracis.	
Minitherm.	37 °C to 110 °C
Bacterium thermos.	

(According to Arloing.)

[28] Mecanique chimique. pp. 89 to 101.

Aspergillus Growth.

A 19 °C C	Ogr.3 harvest
22 °C	O. 6
27 °C	1. 2
29 °C	2. 5
32 °C	3. 5
34 °C	4. 2
37 °C	3. 8
39 °C	3. 0
42–43	Traces.

(According to Raulin.)

NOTE VI.

Thermal classification of some organisms and organs. Order of Maxitherms.

Family I.-50 °C to 60 °C maximum.

Genus I.-55 °C to 60 °C maximum.

Flagellates and zoospores.

GENUS II.-50 °C to 54 °C.

Inf-stentor, human sperm, frog muscles, yeast.

Family II.-40 °C to 49 °C.

Hongo (*Aethalium.*) – Amoeba.- *Actinophrys*. Miliolids. -*Inf. Paramecia.*-Vertebrate Muscles-Many bacteria. *Oscilatoria.*-Nostoc.-*Spirogyra*. Plant cells.

Order of Minitherms.

	Minimum.
Bacillus anthracis	−110 °C
Ciliated frog cells	−90 °C
Ins-bombys eggs (I. L. H.)	−40 °C
Amphibian eggs	−10 °C
Cells of *Tradescantia*	−14 °C

(According to Labbé's table).

Toxic power and specific heat.

	Atomic weight.	Specific heat.	Action.
Lithium	7	—	Weak
Sodium	23	0.2934	”
Potassium	39	0.1695	Energetic
Copper	63	0.0955	Very energetic
Silver	108	0.05701	”
Lead	207	0.03140	”

NOTE VII.

Heat released by one gram of albumin purified	4,998
Muscles of a bull, washed with ether	5,103
Beef Fat	9.069
Hippuric acid	5,383
Uric acid	2,614
Urea	2,206

(Frankland).

Chemical phenomena and temperature.

	To 15 °C	To 500 °C	To 1.000 °C
Hydrogen + Oxygen.	0.	H_2O To 350 °C	← H_2O → To red
Oxygen + mercury.	0.	Oxide.	← O Hg →

SUMMARY:

Thermochemistry can be considered to some extent the basis of biology, but the same thermochemistry is a chapter in the great book of forces, which are fundamentally unified. As such, heat could be the basis of biology, but as heat is simply a form of movement, *biology and all human knowledge can be reduced to the study of movement.*

8.-*The Earth regarded as a living organism.*

In 1900 we presented in the Society "Alzate" a study entitled "Protoplasm and the Cosmos", in which we said that, according to the laws of fundamental unity, the life of worlds is comparable to the life of beings and is characterized by continuous flows and necesary transformations.[29]

In 1902 the important work of Stanislas Meunier was published: "General Geology," which starts with the same comparisons.

Geike says[30] that this world is a living being, like a plant or an animal, however, it would be unlawful to call it that in a way. "The circulation of air and water may well be called the life of the Earth."

Actuality. The doctrine named geology admits that the moment in which we live, in which we study nature, *is not distinguished by any essential geologic character from those which have preceded it and is a simple link in time.*

---

[29] It would be truly amazing if the silicates, the basis of the Earth, were of the animated world, with the same characters, and with identical dynamics and fluid flow.

[30] Concepts in Physical Geography. Veracruz, 1883.

In the ancient texts the world is compared to a creature and is depicted as coming from a seed or an egg. Its incubation, according to the Hindus, lasted three trillion years. The egg of Brahma was divided into two parts, one of them was the sky and the other the Earth.

The first students of geology, Cuvier most of all, had a different idea, assuming the Earth, rather than having developed in a gradual manner, as does an egg, *had suffered tremendous cataclysms* and if there is something certain in geology, it is that the surface of our globe has been the victim of a great and sudden revolution, in an age that cannot go back more than 5,000 or 6,000 years.[31]

Lamarck, since 1802, in his Hydrogeology, had protested against this doctrine, admitting only local disasters such as eruptions, earthquakes, floods, hurricanes, etc.[32]

Hutton said that neither the sign of a beginning nor of an end can be seen in the world.

Hutton based his *theory of the Earth* on the continuous renewal of the continents; Playfair popularized this doctrine and Ch. Lyell, the notable rival of Darwin, was his successor. He acknowledged slow causes.

Actuality and activism.

What C. Prevost formulated is, in short, terrestrial evolution. It is based on facts such as metamorphism and epigenesis or transformation of substance in rocks and minerals. The conjunction of all of these transformations is continuous and not limited, as was pretended.

---

[31] Cuvier was always dominated by mystical ideas. According to Gerard, Cuvier, intelligent, an able analyst, with great sagacity, was nevertheless, for the King's Garden, the Tunic of Nessus. The wanderings of his Protestant education, had inspired aversion to encyclopedic philosophy; he did not understand, from his entrance into the temple of science, all the splendor of this building; he saw only lines to be reproduced, capitals to draw, an architectural arrangement to describe, without deducing a generalization from all of this. He destroyed the great ideals of education, replacing them with the cold data of his Table of the Animal Kingdom. More courtier than naturalist he rose to pinnacle of greatness by his condescension to the wills of the powers that followed him, and consecrated in science a deplorable error, that natural history is the science of details and method is the principal and final object of the naturalist. He did nothing for the triumph of study to emancipate the spirit and impeded the generous men who he made accept his doctrines. Gosse classified Cuvier as Analyticus diplomaticus, Lacepède.

[32] Lamarck (Philosophus clarissimus): who is not moved upon hearing the name of the sage uttered, whose genius was denied and who languished full of bitterness! Blind, poor, rejected, left alone with a glory whose extent was known, but which was only sanctioned as the centuries revealed the laws of the body more clearly. Lamarck, your helplessness, however painful that it was in your old age, is worth more than the ephemeral glory of men whose reputation was due to their association with the errors of his time! Honor to you. I respect your memory. You died in the breach, fighting for truth, and this assures you immortal glory. (Gérard). As we will discuss later, Lamarck was the forerunner of the doctrine of development and his name must always be linked to Geoffroy St. Hilaire and Darwin. A school of naturalists has been founded in the United States, who attach great importance to the environment, and it is called Neo-Lamarckism. It is very uncompromising.

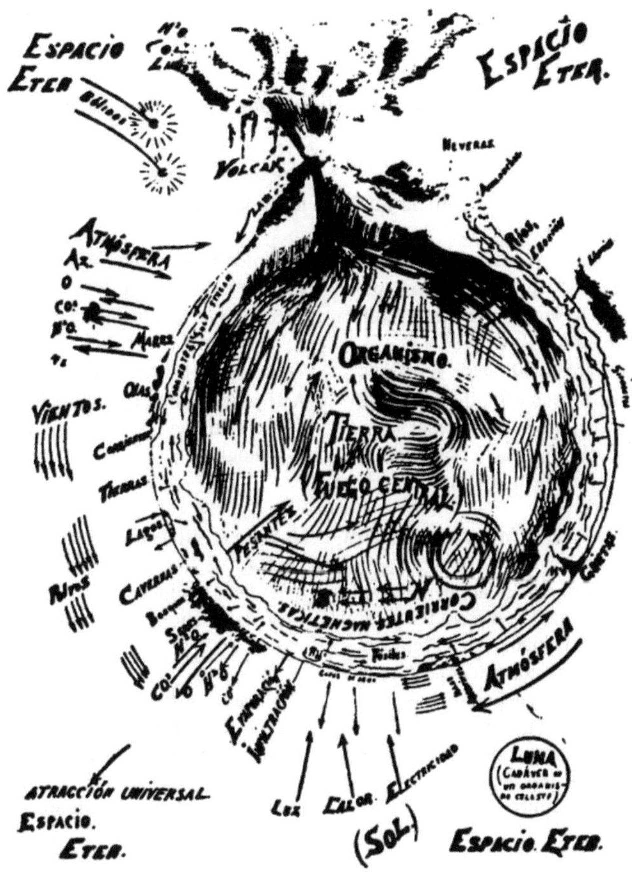

Fig. 9. The Earth seen as a living organism at work. Circulation of matter and force. (Not shown in this schematic figure are the core-area and lacoliths.).

The changes of matter recall those which occur in cells in contact with liquid. In both cases there is a loss and acquisition of materials and the mass gradually exchange its initial molecules for new molecules.

Nutritive currents are a condition of life is the whole can be compared to a river and a field that it fertilizes.

***

A Jurassic layer can be formed by elements of very different ages. A ferruginous layer may acquire its iron long after the initial period of formation.

Meunier speaks of a telluric physiology and mentions the following facts:

Moving equilibrium of the medium.

Phenomena due to high temperatures, heaviness, heat and sunlight.
Purely mechanical phenomena.
Purely chemical phenomena.
Heliogenic phenomena whose immediate cause lies in the activity of organisms.

Telluric organism.	*Animal and plant organisms.*
Constant current, circulation of nutrient fluids, rivers, oceans, groundwater.	Id. Blood. Saliva.
Less nourished parts, crystallized terrain.	Id. Epidermis, woody, bones.
More nourished parts, banks, deltas, coasts.	Tissues in formation.
Breathing, oxygen absorption by water, solfataras, production of CO_2, etc.	Breathing. The function of chlorophyll.
Internal heat, physiological factor. Lacoliths.	Thermogenesis. Animal heat. Flowering.
Activates streams of water (Gulfstream) and various materials; produces thermal springs, soffioni, fumeroles; seems to determine almost all the phenomena of planetary physiology (Meunier).	Activates osmotic flows, breathing. Thermochemistry can be considered as the basis of biology.
Heaviness produces flows in water, sand, lava, etc.; infiltrations.	Heaviness influences geotropism, geotactism, etc.[1]
Solar activity causes geological phenomena like rain. Earth still retains heat from the primitive solar mass. There is inequality of temperature, wind, dust. Equator caldera. Polar coil.	Solar heat has great influence on life; chlorophyll function, transformation of radiation.
These can be regarded as apparatuses of geological activity: flexible cortex of the globe, the volcano, the layer of groundwater, deep or shallow, the sea, the snowstrom, the atmosphere and living beings.	There are similar devices, circulation, respiration. A parallel could be established by comparing bones to mountain ranges, rivers to arteries, the tracheae to the vents, the layers of the cell membrane to the strata, the calcareous deposits of the vessels to those of a hot spring.

[1] Labbé. Cytologie expérimentale, p. 81.

Continents migrate on the surface of the Earth (Huxley).

The *flexible cortex* protects the outer regions from internal heat, is modeled on the nucleus, which is every day less, forming gradients, which influence other phenomena of planetary physiology. The *volcano* is a communication between the atmosphere and the interior regions of the globe, and consequently, one of the most active centers of circulatory phenomena, in particular of water, which circulates vertically through a great thickness of crust.

The deep water layer, the system of cavities of all kinds, wells and caves, resembles the circulatory system of plants and animals, the volume of water associated with rocks is higher than that of the seas, this water circulates and gives an appearance of life to the deep geological environment.

Finally this water is absorbed by the rocks and the same has happened on the moon, the skeleton of a celestial organism. This movement produces CO_2 and other gases. *The surface water layer* is linked to the preceding geological structures. *The sea* is a gigantic laboratory for demolition and sedimentation. It decomposes rocks, precipitates manganese oxide, and regulates the amount of carbon dioxide and water vapor in the atmosphere. The *icy region* is an agent for continuous water circulation, transport of rocks, and denudation. The *atmosphere* is an agent of denudation, sedimentation, chemistry, etc. A living being is a single organ for denudation and sedimentation.

All these apparatuses have worked in every moment of the life of the globe since it began to cool.

9.-*The unity of substance in the organic and the inorganic world.*

The universe is a completeness which is eternally mutable in form, forever unchanging in is foundation.[33]

Living or organized bodies cannot be made of a special substance. They are an integral part of their environment, and only leave it to return to it. There is not an atom of their substance that does not participate in the eternity of universal matter, the basis of all that exists.

There is not an atom that does not have an infinite number of destinies in endless organic or inorganic combinations and is not called to many changes in its association with other atoms.

The human body is composed of C, H, O, N and salts, whose elements have operated in very different circumstances, in volcanoes, in tombs, at the bottom of the sea and clouds, in bacteria, in space, in meteorites. The history of each of these atoms would occupy entire libraries and their future is an infinite extension.

Shakespeare said that the ashes of Alexander were perhaps serving to prevent air flow through a filthy crack.

Elemental analysis of the most complex body of animals, man, reveals, mainly, fourteen bodies of mineral chemistry:

1. Oxygen.	6. *Phosphorus.*	11. *Calcium.*
2. Hydrogen.	7. Fluorine.	12. *Magnesium.*
3. Nitrogen.	8. Chlorine.	13. *Silicon*
4. Carbon.	9. Sodium.	14. Iron.
5. Sulfur.	10. Potassium.	

There is no radical difference, no specified limit between organic and inorganic chemistry: both study the same basic bodies, subject to the same laws. Organic substances come from inorganic ones and return to them incessantly, only to leave them again. Except organic substances usually have a greater complexity. Little by little, modern chemistry disputes and snatches the monopoly on the manufacture of organic materials from living bodies.

On the other hand, if minerals and organic compounds are placed in a graded series, one finds that between the two are transient groups, serving as links: the carbides of hydrogen, carbonic acid, ammonium chloride, ammonium carbonate, alcohols, ethers, ternary acids, fatty substances, whose synthesis is carried out in

[33] Letourneau. Biologie, p. 11.

the laboratory. The Sesquicarbonate of ammonia, for example, has the following formula:

$$(CO_3)_3(NH_4)_4H_2 + 2\,H_2O.$$

With this formula another can be obtained, by arranging the atoms in a different way:

$$C_3H_{22}N_4O_{11}$$

or multiplying by 10:

$$C_{30}H_{220}N_{40}O_{110}$$

This formula looks similar to that which Schützenberger gives to albumin:

$$C_{60}H_{100}N_{16}O_{20}$$

There is nothing fatal in the composition of organic products. The chicken egg shell calcium can be replaced by magnesium.

In fats the hydrogen can be replaced by chlorine, without substantially changing the properties of the compound. Organic substances have been able to be reproduced by synthesis. Fischer has developed a kind of albúmina.[34]

Cyanide can be prepared by passing nitrogen, at high temperature, over carbon impregnated with potassium, and ammonia from nitrogen and hydrogen. The synthesis of complex organic matter, using simple principles, has reduced the general problem to the formation of a small number of fundamental principles: the synthesis of alcohol was reduced to that of ethylene C_4H_4 and this to C_4H_2 or to that of acetylene. Finally, Berthelot synthesized acetylene from a stream of hydrogen, which passed in a glass flask which produced, between two cones of coal, a voltaic arc of a battery of 50 elements.

***

It was reported that the albuminoid substances were the basis life, but we doubt this claim, because we can only consider them as fuel or reserves.

It has also been said that these substances are distinguished primarily from the inorganic ones because they could not crystallize, but Wroblewsky and Hofmeister have recently obtained different albumins which crystallized perfectly, and there are uncrystallizable inorganic substances, such as certain phosphates and silicates.

In the end, it was claimed that the essential organic bodies were colloids, that is that they did not pass through osmotic filters, but colloidal metals have been obtained with the same properties. We will explore this later in this chapter, when we discuss the ferments.

***

[34] Grimaux. Synthèse des principles azotés de l'organisme. Journal of l'Anatomie et de la Physiologie, 1880, nos. 2 and 4.

All organic substances are composed of carbon, hydrogen, oxygen, nitrogen and salts, namely inorganic bodies.

We have proposed, finally, a tentative hypothesis: that the protoplasm has as its base inorganic colloidal silicates and phosphates, secreting them or accumulating organic principles, which would be secondary and not structural.

In this concept, biology and geology should be twin sisters, like inorganic and organic chemistry.

***

Iron, which forms the slave's chain, may, over time, become iron oxide, which falls to the ground and forms a sulfate or nitrate, and then passes into the plant and then the animal, forming an essential part of blood, which gives a rosy tint to the cheeks of a young woman.

Under the influence of a blow, a poison or any other accident, animate matter begins to suffer further transformations. Subjected to the action of insects, bacteria, or evaporation it becomes an inorganic residue, completely inorganic. Mineralization of sugar requires the intervention of two ferments: yeast, an anaerobic organism, converts it into alcohol and carbonic acid, *Mycoderma acetii*, an aerobe, burns alcohol and leaves in its place CO_2 and H_2O.

Ferments, according to Duclaux, convert all of the materials of the organism to gas, but as the transformation is incomplete, its products are at once in the ambient atmosphere and the fermentable medium.

***

The circulation of the matter is an undeniable fact, and organisms are very complicated and efficient apparatuses for that movement.

It is clear that if organic matter were not broken after death, it would drain the life from the Earth. That is, atoms and inorganic molecules are *loaned* to living beings while they live and returned to nature when they die.

In the former scenario, referring to the appearance of the living organic substances under mysterious conditions, and forming by themselves, it is not clear how this movement may be necessary for matter.

***

Parasites of the dead, we can say that upon hosting a banquet, are preparing the dishes for the following one, as they release inorganic matter, so necessary for life. It is a kind of commerce, which is based on an unchanged capital, which circulates endlessly.

***

Circulation of matter

We have said that, according to Preyer, there is an almost constant relationship between the mass of living matter and the mass of dead matter, and this could be no other way, taking into account to the above considerations.

9 a.-*Living beings considered as colloidal minerals. Provisional insights. Provisional hypothesis.*[35]

Several years of experimental studies regarding the imitation of protoplasm have led to an interim theory of organization, we do not present it as demonstrated truth and it should be regarded with extreme distrust, until it is subjected to a thorough examination by authoritative researchers.

After testing a multitude of reagents, continuing the work of Bütschli, who prepared a kind of protoplasm with oil and salt or alkali; after testing a thousand different albuminous subtances, in which could never detect a structure or ameboid movement, we observed this interesting phenomenon: metaphosphoric acid produces microscopic nipples in egg white after addition of calcium chloride, which slowly deform and have very apparent internal movements.

Metaphosphoric acid triturated with calcium carbonate or chloride, also produces a multitude of drops sarcodicas or ameboid masses, which are very similar to natural ones.

Moreover, Leduc obtained cells with nuclei using copper ferrocyanide.

Both of these facts, along with others described in dealing with the cell, led us to suppose that the essential matter of beings is an inorganic, viscous, colloidal salt, functioning as an osmotic electrolithic apparatus, dissociating ions, secreting albumins (combustible bodies) and various reserves.

Whatever the merit of this theory, which is not for us to judge, we are certain that for now it is the only one able to explain the emergence of living beings, their dependence on the inorganic world and the formation of a new protoplasm in a developing organism, under current conditions on Earth.

If we agree with biologists in general that the basis of living beings is proteinaceous substance, we cannot understand that this is formed by some cause X formed in early geological ages, persisting, after millions of years, and formed itself today:

"*Albumins derive from albumins. Albumins form albumins*. This means that John is the father of John, the son is the father of his father, sulfuric acid forms sulfuric acid, water forms water, nitrogen forms nitrogen." This is absurd.

We assume the contrary, that John forms Peter, that sulfur and oxygen form sulfuric acid, that inorganic phosphates or silicates absorbed by the organism form organic substances, more or less complicated and unstable. (See protoplasm).

***

Instead of a metaphosphate or inorganic silicate may not the basis of Life be an ammonium salt? We think not, because there is no insoluble salt except the picrate, the chloroplatinate and others that have been made in the laboratory. Ammonium phosphates are soluble.

---

[35] We adopt here a prudent tautology, as we are not sure of the accuracy of our ideas.

Soluble ammonia salts

NH_4Cl, NH_4Br, NH_4I, NH_4F, NH_4SiO_4, NH_4CN, Sulfohydrate, Sulfocarbonate, Nitrate, Nitrite, Chlorate, Chlorite, Bromato, Sulfate. Sulfite, phosphates, Hyposulfate, Selenite, Tellurte, carbonate, cyanate, Phosphates, neutral and biphosphates, phosphite, Hypophosphite, arsenate, arsenite, Borate, anhydrous sulfate, anhydrous Sulphite, Carbamite. Magnesium ammonium phosphate is slightly soluble in pure water.

Insoluble ammonium salts

Bitartrate, ammonia-alumcal, phosphomolybdate. Cloroplatinate, magnesium ammonium phosphate is insoluble in water containing salts. Ammonium picrate, phosphates of calcium and ammonia.

Furthermore, ammonium salts decompose even in the cold, by fixed alkalis and liberate ammonia. The protoplasm contains much calcium (plasmodia).

***

How was the first particle of albumin formed if there was no other previous one, to produce the necessary elements?

It would be very logical to assume that an inorganic emulsion works as an osmotic device that combines the elements of air and water, forming many compounds[36]. In fact, the air, a mixture of O, N and CO_2 water (H_2O) contains all of the elements of albumin and thus theoretically would form (scheme):

Carbonic acid from the air.............................60
Water..50
Nitrogen from the air or from nitrates............16
That is:

$$C_{60}H_{100}N_{16}O_{20},$$

which is the formula of albumin, according to Schützenberger, and a residue of 152 oxygen. *However, plants give off this gas in considerable quantity, under the influence of light!*

(Actually the amount of excess oxygen is much higher, since the air contains only 4 to 6 parts in 10,000 of carbonic acid).

Starch is formed from

CO_2 18 parts.
H_2O 15″

or

$$C_{18}H_{30}O_{15} + 36 \text{ of oxygen.}$$

(Starch).

***

[36] Beijerinck and Van Velder have studied some microbes (*Bacillus oligocarbophilus*) which take volatile carbon and nitrogen from the air, and are thus "biological purifiers." (Versl. E. Akad. Amsterdam. DIXI, 1902-3, p. 460–465).

It should be made clear we do not claim to have discovered the mechanism of formation of protein and starchy substances, because we are only concerned about the idea of finding a simple explanation of the phenomenon, which is for now hypothetical[37].

The oxygen absorbed by all living protoplasm, animal or plant, is used for combustion. For example, acetic acid ($C_2H_4O_2$), under the influence of oxygen and *Mycoderma aceti* turns into water and carbonic acid:

$$C_2H_4O_2 + O_4 = 2\,CO_2 + 2\,H_2O$$

Now, under the influence of the ferment of Friedel

$$2\,CO_2 + 2\,H_2O \times 9$$

could result in a plant:

$$\underset{\text{(Starch)}}{C_{18}H_{30}O_{11}O_{40}} + 3\,H_2O \rightarrow$$

or regenerate the acetic acid that exist in plants:

$$9(C_2H_4O_2) + O_{36} \rightarrow$$

Dehérain[38] says that cells with chlorophyll exposed to the sun break down CO_2 and emit a proportional amount of oxygen and at the same time (*du même coup*) is full of starch, "although they proposed some intermediate reactions that are not proven."

***

Despite what we have said it is understood that the theory of life is yet to be discovered, that our hypothesis of inorganic protoplasm lies in a small and insufficient number of experiments and that the reasons given in their favor can be misleading, since chemical biodiversity is still not sufficiently clear and detailed and the protoplasm is composed of 30 or 40 very different substances, whose functions not been elucidated yet.

Tomorrow perhaps it will be proven that proteinaceous substances are easily organized and form osmotic structures, such as oleate or tannate, and that their origin is explained in thesimplest manner, by the action of electricity on nitrates, or any other natural process.

We suggest these doubts to students, hoping that they maintain a philosophical doubt and disbelieve in the certainty of this or that premature explanation. *Provisionally we assume that the basis of life is an inorganic emulsion, while studying the issue thoroughly.*

---

[37] Verworn attributes life to a kind of ferment, the biogen, which is completely hypothetical. Die Biogenhypothese. Eine Kritische-experimentelle Studie über die Vorgang in der lebendige Substanz. Jena. Fischer, 1903. 8th., pp. 1–114.

[38] Origine du carbone des végétaux. Revue Scientifique. 18 Novembre 1880, p. 469

9 b. *Parallel between living beings and minerals.*

1. Living beings and inorganic bodies are material, they occupy space, react under the influence of physico-chemical agents, are made up of infinitely small particles, atoms and molecules, and have a variable form.

2. There are colloidal and non-colloidal bodies.

The former can be inorganic and yet have a sort of rudimentary organization. For example, artificial cells and emulsions of colloidal calcium phosphate, potassium ferrocyanide, of calcium saccharate and sodium sulphate, of silicates and colloidal silicates.

There are similar emulsions, oleate and tannates, which are not living, but they are of mixed nature (organic acid and inorganic base) such as sodium oleate, or completely organic, such as the tannate of gelatin.[39]

3. Organic or inorganic colloidal bodies show an instability that was previously thought to be a feature of life but it is not really. The emulsion or colloidal state makes these bodies very sensitive to the action of reagents and they are easily disaggregated or dissociated.

4. As we have said, life was probably had an inorganic, mineral origin, and it maintains itself, after millions of years, on the same basis, perhaps an inorganic silicate, which certainly exists in nature in great abundance, forming huge deposits.

5. The crystallization of many inorganic salts resembles biological phenomena and a glass that regenerates the missing parts, also has a constant form and is susceptible to increases (by juxtaposition).

6. Many organisms, such as calcareous sponges and foraminifera, have regular shapes. The same is seen in the hairs of many plants.

Rainey's experiments show that the carbonate of lime, crystallized within a colloidal body can form concentric patterns.[40]

7. Colloidal organic matter cannot live without inorganic matter.[41]

Even albuminous or proteinaceous materials are always linked to various inorganic bodies such as sulfur or phosphoric acid.

8. It had been assumed that life was characterized by uncrystallizable substances, but Hofmeister, Wroblewsky, and Ruempler[42] have shown that these bodies can crystallize. We have obtained a sample of albumin, crystallized by the procedure of Wroblewsky sent to us by G. Serono, from Italy.

9. Another difference was in the immortality of minerals, but we have seen that everything in nature is in motion and even iron undergoes molecular changes that make it brittle. Large trees are killed by accident. Protozoa, which divide and multiply by dividing, are immortal, according to Weismann. The rotifers, tardigrades and other revivable animals, if they are dry, do not die, in an indefinite time. Chlorine solutions are very changeable, and so on.

---

[39] It seems that tannic and oleic acids contain some inorganic impurities, perhaps alkali silicates.

[40] Carpenter. The Microscope, p. 842.

[41] Loew. The mineral nutrients. U.S. Department of Agriculture. Washington, 1901.

[42] Pharmaceutical Journal. London. April 20–1901.

10. Approximately 50,000 carbon compounds are known, whose study is the subject of organic chemistry,[43] but for the most part are synthetic products. Even nuclein or nuclear albumins are composed of inorganic elements:

C, H, O, N, P,

the same as any of the 50,000 organic bodies, and are based hydrogen carbides. We have already said, that these could be formed simply from air and water.

11. Mention has been made of real seed crystals. In a study of M. Gerncz presented by Pasteur in the Paris Academy of Sciences (meeting of June 13, 1874) it was noted that in a solution of sulfur in gasoline the formation of octahedral or prismatic sulfur can be determined by adding a microscopic crystal of octahedral or prismatic sulfur. There are other very suggestive facts: glycerin crystallizes under as yet undetermined conditions, that is, it acquires a structure similar to protoplasm.[44]

Parallel between the inorganic bodies and living beings.

Living beings: material, occupy space, react with different agents, have a form, are composed of atoms.

Bodies: material, occupy space, react with different agents, have a form, are composed of atoms.

Living beings: are composed of organized, unstable colloids.

Bodies: sometimes consist of colloids (phosphates) and are unstable (colisilicates).

Beings: present phenomena of regeneration, growth, healing, special form.

Bodies: crystals[45] present phenomena of regeneration, growth, healing, and special form.

Beings: usually have symmetric and geometric shapes. (Living plants, protozoa, epidermal productions).

Bodies, often have symmetric and geometric shapes. (Crystals). Some are amorphous.

Beings: based on inorganic substances: P, Si, Ca, S, Fe, etc., and crystallizable albuminous compounds.

Bodies: are based on inorganic substances, many of them crystallize.

Beings: are sometimes regarded as immortal (protozoans, dried rotifers). Others have finite lives (humans).

Bodies: appear to be immortal, but everything in nature is in the process of constant transformation. Some inorganic bodies are unstable (hydrogen peroxide, ferric chloride of phosphorus, etc.).

Beings: feed themselves.

Bodies: the cells of copper ferrocyanide show a similar phenomenon. (Leduc).

---

[43] Science. XV, no. 391, p. 1,016.

[44] See: Dastre. La vie et la mort. Paris, 1903.

[45] See, regarding the study of external and internal generative forces in crystals and the body: Haeckel. Histoire de la Création, p. 297.

9 c. *Arquigonia. Spontaneous generation. Bathybius, Protobathybius, Eozoon, Colloidal phosphate. Clay.*

There is a very general and old (as old as the world) belief: the spontaneous generation of living beings. The most distinguished scholars of ancient and modern times have believed in this. Aristotle attributed the generation of eels to a sort of spontaneous decomposition of silt. Van Helmont seriously attributed the birth of mice to the decomposition of a piece of cheese wrapped in a dirty rag.

Belief in spontaneous generation reigned undisputed in science at the beginning of the seventeenth century. The experiments of Pasteur and Tyndall showed that in laboratories all life comes from preexisting life, whose seed is in the air, water, etc., and to prove this it is sufficient to note that sterilized milk and canned food heated and stored in metallic boxes remain unfermented indefinitely.

***

Spontaneous generation possibly took place and may still take place in saline soils or in brackish or mineral-containing water. There really is no reason to look for it in boiled aseptic organic liquids or organic substances, where every hydrosol or emulsion has been previously destroyed.[46]

Let us suppose that life appeared in water, forming albuminous substances at the expense of ammonium nitrite in the atmosphere or of a hypothetical compound, perhaps cyanogen, or if you will, by the evolution of a living germ carried by a meteorite (Thomson's theory). However, neither nitrite nor cyanogen, or that meteoritic germ are found in the body fluids or other substances that Pasteur and Tyndall investigated.

Haeckel said[47] that the impossibility of spontaneous generation cannot be established.

In addition, in our time, before our eyes, under the current conditions on Earth, millions of cells and tons of proteins and protoplasm are forming through the inorganic elements that plants absorb from the air, water and earth.

How do we conceive that the first germs of life, born under conditions very different from those of today, have preserved their way of life and produce the same results after 30 or 60 million years? And that they live in very cold or very hot, very wet or very dry conditions, with no air (yeast) or in the air, to 4,735 meters of depth[48] or in Tibet, more than 5,000 m above sea level.

That is why Schaaffhausen boldly admits that water, air and mineral substances are combined directly under the influence of light and heat, giving rise to a colorless Protococcus.The supposed genus *Eozoon* (Fig. 10) (the dawn of life) was discovered by Logan, of the Geological Survey of Canada, in 1865, and Dawson gave this name. There has been a long discussion about this issue. Some say it is the oldest

---

[46] And even though these these conditions are presented, they would advance very, very little, as the same question remains: how was the organic medium required for the spontaneous generation of organic life formed?

[47] Histoire de la Création, p. 300.

[48] *Mol-fusus abyssomm* (I. G.) 2 Delage. La structure du protoplasma, p. 402.

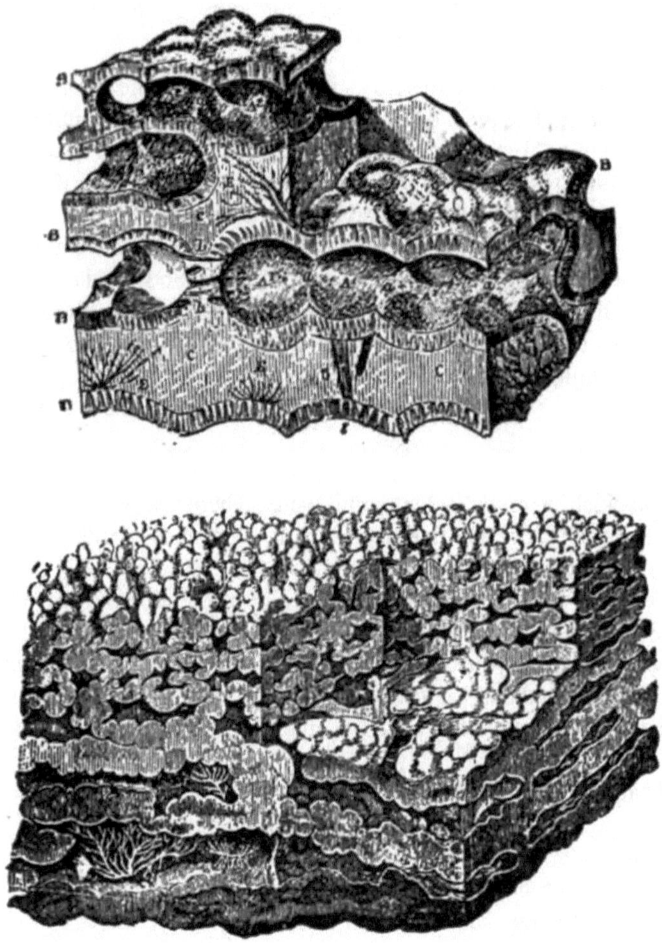

Fig. 10. Eozoon canadense. Nests of some hypothetical primordial organism, also considered to beaccidental formations of mineral and metamorphic origin.

foraminiferan and others that it is entirely inorganic. The first notion was held by Carpenter and Dawson, and the latter by most petrologists and geologists, starting with King and Rowney and later by Möbius. Bütschli granted some support the former, but is more inclined towards the other opinion. Petrologists claim that same structures are seen in undoubtedly igneous minerals.[49]

Eozoon is found in limestone layers of serpentine in Laurentian land base of Canada (part of the Archean, the oldest in the primary). It looks like a coral reef. It form irregular masses composed of calcium carbonate and magnesium silicate, with the appearance of real shells with irregular chambers and ramified channels.[50]

[49] Calkins. The Protozoa, p. 29, note.

[50] Carpenter. The Microscope. p. 587. *Eozoon canadense* was called by some light spirits, "Eozoon Canardense", from *canard*, sensational and baseless news.

Fig. 11. *Bathybius Haeckelii*, the supposedly primordial organism found in the great depths of the ocean by naturalists aboard the steamship Challenger.

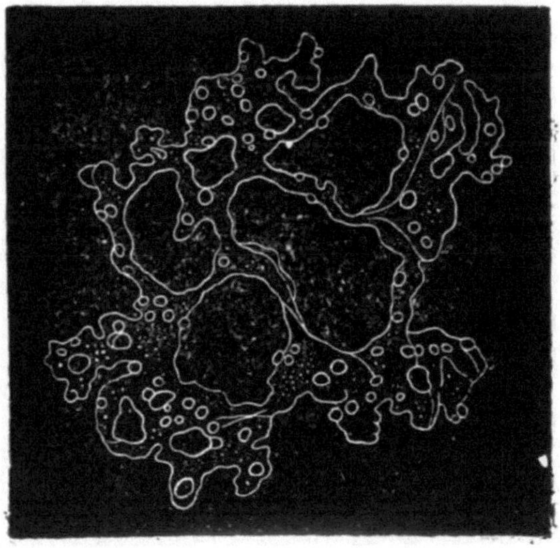

***

Bathybius Haeckelii (Fig. 11) was found by the naturalists of the steamship Challenger, between 4,000 and 8,000 m deep in the sea. It is a kind of protoplasmic network, considered by some to be a calcareous precipitate or the remains of sponge larvae[51] (Fig. 11).

Protobathybius was collected by Bessel in the polar seas. Its discoverer said that this protoplasmic mass that has shown "magnifient ameboid movements."[52]

In summary, nothing certain is known about the origin of life and its earliest forms.

Undoubtedly they were aquatic, and as salt is the medium in which all living cells of almost all organisms live, it is reasonable that they appeared in the sea or saline soils.

We assume that an inorganic silicate was found in these conditions and began to live, but this hypothesis could be entirely false.[53]

***

The discussions and experiments of Pasteur and Pouchet have had this idea as their basis: is spontaneous generation possible in culture media in laboratories, or in organic media, in broth, or in eggs?

This way of presenting the problem has always been sophisticated, because nothing is said about the origin of such organic media. If spontaneous generation occurred in them there would be this serious objection: if they are essential for the appearance of life, how were they formed when there was not yet and life or organization?

---

[51] Haeckel. Histoire de la Création, p. 166.

[52] Perrier. Les origines of la vie. "Nature." 8 March 1879, p. 210.

[53] A. L. Herrera. Le protoplasma de métaphosphate de chaux. 1902.

***

When we were preparing the manuscript of this book for delivery to the typographer, we were in the process of studying an absolutely positive explanation which in some respects will satisfy the thinker who exercises his intelligence in generalization of all the facts of nature by the light of a doctrine of universal unity and coordination.

This theory is simply to suppose that silica and colloidal silicates (clay), which are abundant in the sea, on land and in organisms, are the cement of protoplasm, its structural basis, the indispensable osmotic apparatus. Indeed, in the laboratory, they produce a multitude of almost organized structures and exhibit nearly physiological properties (see imitation of protoplasm).

But yet this hypothesis has not entered the category of incontrovertible scientific principle.

Meanwhile a parallel explanation can be accepted as true and is very closely related to Preyer's law of conservation of life:

***

Law of conservation of life.

"Living masses, according to Preyer, are in an almost constant relation with the mass of dead matter.

"And yet, many naturalists have written that spontaneous generation, which now appears to be untrue, was able to populate the globe in early geological times. Is there a less scientific conception? If this were so, a capital law of biology would have become dormant! This would be equivalent to accepting the relationship of Genesis with its primitive candor.

"In this way the opponents of naturalists have rightfully mocked them.

"If biology is an exact science it must be admitted that its laws are immutable and are now what they have always been.

"Physics is independent of history, its laws do not contain any element that depends on time. It is now the case that biology is established as an exact science.

"The eternal nature of life is the consequence of the Preyer's important law about the constancy of the mass of plasma, that is, that biological laws have the same character as the laws of physics, they are absolute, independent of any notion of origin and end, all consideration of time and place, and therefore, biology is an exact science."[54]

Indeed, as we said at the beginning of these lessons, lime, phosphoric acid, iron and other elements of organized beings are found in extra-terrestrial bodies and the general biogenetic law should be expressed throughout the universe.

At this time no one can prove that there is no spontaneous generation in the saline waters, in the sea, or in springs.

---

[54] G. Sorel. La loi de la conservation de la vie. Revue Scientifique, 28 °C année (2) p. 276.

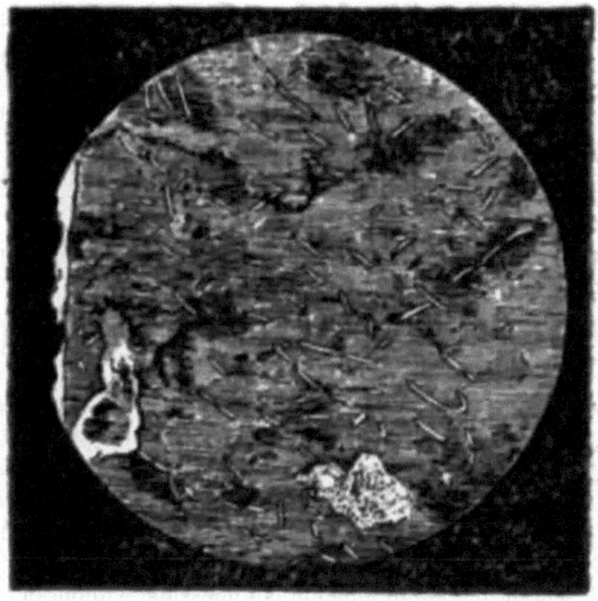

Fig. 12. *Bacillus permiensis* in coal, much similar to today's microbes. After Coupin. "Nature". 1902 (1) p. 45.

It has been sought in precisely those places where it should never take place: under artificial conditions, not under natural ones or imitations of them.

Moreover, Meunier justiably says that the Trilobites were certainly not products of a very different environment than today's, because their organization does not differ much from that of modern crustaceans.[55] Van Tieghem has found the butyric microbe (*Bacillus amylobacter*) in rootlets of conifers, during the era of coal, 40 or 50 million years ago." (Fig. 12).

Incidental to theories of life, we say that the biologists have been concerned, so it seems, considering these substances as the basis for the organization of the protoplasm or tissue, which cannot be accepted for several reasons:

9 d. *The dogma of albuminous substances.*

1. They dissolve in salt water.
2. They would form by themselves and the appearance of the first particle of albumin is not conceivable.
3. In laboratories they do not form structures when they have no inorganic impurities.
4. They would be destroyed by the enzymes and general components in protoplasm.
5. They are easily transformed into lecithin (a phosphorylated fat in salmon, according to Miescher) and lecithin is regarded as a necessary food for the brain.
6. They are oxidized and converted into urea. Therefore, a fully active protoplasm would destroy them quickly.

[55] C. R. Acad. Sci Paris, 29 dec. 1879. 3 General Geology, p. 317.

7. They are forming on Earth under current conditions, especially in plants.[56]
8. They are proposed as the origin of life though their extraordinary complexity is recognized, which is incompatible with this fundamental creative role.
9. They are not found in nature outside organisms or independently of them, they have not been formed under favorable conditions, under the influence of light, in unstructured media (the experiments of Tyndall and Pasteur).[57]

9 e-*Dominance of mineral substances in biological phenomena.*

Clearly, these are the main tests:

1. Inorganic, protoplasmic, artificial structures, endowed with certain physiological properties.
2. Intimate and necessary relationships between plants and the inorganic medium. The giant Baobab, is fed only with air, water and salts.
3. The organism is a combination of organic and inorganic bodies. The albumins contain sulfur or phosphorus. The enzymes owe their activity to iron, manganese (oxidases), etc. Inorganic ferments have been prepared, such as colloidal platinum.
4. The formation of albumin is impossible without phosphates. Animals fed with pure organic materials without mineral ingredients die quickly.
5. Salt, calcium, magnesium, silicon, iron, manganese, are essential for life, as is water.
6. Salt, mercury and other drugs, have a strong and often lasting action.
7. The electrical conductivity of organisms is due to their water and salt contents.[58]

10. *Parallels between plants and animals.*[59]

There are no fundamental differences between the plant and the animal kingdom. The protoplasm, which is actually living, is very similar in both kingdoms, in its structure and functions, but has adapted to different media, defended by various mechanisms from drying, cooling, mechanical causes of disorganization, etc.

[56] See the discussion of this issue: A. L. Herrera. Le rôle des substances albuminoides du protoplasma. Revue Scientifique. 1908. (June 10) p. 46. It is notable that nucleins or phosphorus albumin of the nucleus only slowly dissolve in alkali, like the silicates, and how they resist the acidic or neutral solvents.

[57] A bottle of water charged with carbonic acid, ammonia and sodium phosphate was placed in the sun for many days: we have not observed the formation of any body or agency.

[58] A. L. Herrera. Le rôle preponderant des Substances minérales dans les phénomènes biologiques. Mémoires Soc. Sci. "A. Alzate." Vol. XIII, pp. 338–348. 1903. Revue Scientifique. 13 June 1903, Bulletin de la Société Mycologique de France. T. XIX, 3e. fasc. 1903.

[59] Huxley. La zone frontière entre les animaux et les vegetaux. "Revue Scientifique." 8 July 1876, p. 26. This refers to an important link, a lower organism intermediate between plants and animals, Monas lens.

1st-General shape and organization.[60]

Plants.	*Animals.*
External development.	Id. Internal.
Branched forms.	Id. non-branched.

Links.

Polyps and seem Siphonophores resemble plants. (Coral, Gorgons).
Protists. Protozoa.

2nd-Tissues.

Plants.	*Animals.*
The cells are usually separated by walls of cellulose.	Generally fused.

Links.

Cartilaginous cells, especially those of certain zoophytes have partitions.

3rd-Reproduction

Similar in plants and animals. Spores, sperm, oogonia, ova, are fundamentally equal.[61]

4th-Chemical composition and molecular processes of change.

Plants.	*Animals.*
Predominantly ternary bodies.	Nitrogenous bodies
Chlorophyll.	Cholesterol.
	Lecithin.
Cellulose.	Urea.

[60] Claus. Zoologie. 1884, pp. 6 to 13.

[61] Geddes et Thomson. L'évolution du sexe, p. 166.

Fig. 13. A carnivorous plant. *Cynananchum Macrorhizon* Ribesiaceas.

Links.

Protoplasma, identical to that of animals.

Carbohydrates. Fat. Sugar. Glycogen. Chlorophyll (*Stentor, Hydra, Bonellia*).

Cellulose in salps.

Fibrin, vegetable casein, albumin.
Fungi lacking chlorophyll.
Cholesterol in legumes.
Lecithin in seeds.
Urea in fungi.[a]

[a] Bamberg M. und A. Landsiedl. Ueber ein Vorkommen von Harnstoff im Pflanzenreiche. (Sitzungsber. K. Ak. Wien. 1903, p. 44).

5th-vacuoles.

Plants.	*Animals.*
	Contractile vacuoles.

Links.

Contractile vacuoles also exist in plants, according to Maupas.[62]

6th- Chemical function.

Plants.	*Animals.*
Absorb water and salts and form organic compounds.	Absorb water, salts and complex plant products.
Produce oxygen.	Absorb oxygen.
Synthesize.	Analyze.
Accumulate forces.	Transform them into living forces.

[62] Comptes-Rendus Acad. Sci Paris, 19 June 1876.

Links.

Carnivorous plants (Fig. 13). Saprophytic fungi. Respiration is the same in plants and animals and consists of the absorption of oxygen and the emission of carbon dioxide. Ferments and microbes in general absorb nitrogenous combinations and do not assimilate CO_2 from the air, instead taking up carbon from carbohydrates.	Oxygen accumulates in the swim bladder of fish.

7th-Voluntary movement and sensation.

Plants.	*Animals.*
Immobile.	Moving.
Insensitive.	Sensitive.

Links.

Myxomycetes. Zoospores. Sensitive.	Motile.	Polyps. Very degenerate and sedentary parasites.
Spermatia. *Hedysarum.* Stamens and styles of various plants.	Irritable.	

Some authors say that plants are sensitive. And in fact, the protoplasmic masses react in a similar manner under the influence of shocks, but the reactions of plants do not occur by means as delicate as those of a higher animal.

Haeckel's kingdom of Protista includes intermediate bodies between the plants and animals, but has not been accepted.

Nevertheless, Huxley established an interesting parallel between certain flagellated protozoa and inferior algae.[63]

***

Chlorophyll has been considered as one of the essential principles of plants, but Oersted has observed that some of the lower plants may or may not this substance:

With chlorophyll.	Without chlorophyll.	With chlorophyll.	Without chlorophyll.
Oscillaria.	*Beggiatoa.*	*Spirolina.*	*Spirochaeta.*
Leptothrix.	*Leptomitus.*	*Palmellaceae.*	*Chroococcaceae.*
Chlamydomonas.	*Chlamydomonas hyalinus.*	*Synedra*	*Sinedra putrida.*

[63] Huxley. On the Border Territory Between the Animal and the Vegetable Kingdoms. Collected Essays. Vol 8, 1896.

11.-Fusion of zoology, botany and mineralogy.

In the not too distant future these three sciences should be merged, forming part of Geology or part of the description and study of the Earth:

$$\textit{Geology.}\left\{\begin{array}{l}\textit{Mineralogy.}\\ \textit{Botany.}\\ \textit{Zoology.}\end{array}\right.$$

Or better in this way:

$$\textit{Geology}\left\{\textit{Mineralogy}\left\{\begin{array}{l}\textit{Zoology}\\ \textit{Botany}\end{array}\right.\right.$$

The influence that this merger would exercise on the progress of science will be, without a doubt, extraordinary.

Geology will acquire a new importance, and philosophies will need to head in an entirely new direction, considering the reason for life on planets and not primarily the cause of our small biological and social problems.

We began by studying what we believe now and in general about all species and all living organisms. Later, this will preferably be believed about everything contained in the universe, *everything that moves, worlds, protozoa and atoms*.

In this way the *lower* sciences, let's say histology, comparative anatomy, phytography, geography, zoology and botany, will lead to a *higher* science, the Darwinian or transformationist biology, and this, to a *supreme* science, general Cosmology or the science of the universe.

Beyond the mists of the unknown, there must be an order of things related to general cosmology. Perhaps this chain, whose links appear to be becoming simpler and larger, can be extend even further.

12.- The unitary plan of nature.

The only plan that fundamentally has followed nature has been this: to determine the relentless transformation of force by ever more complicated means.

It could be said that the Cosmos is an organism in which circulates a sap, an ether, a blood, an inexhaustible nourishment for the vortices of each internal material particle.

But this food, like that of humans, does not have to be swallowed and discarded without change, it is modified and transformed.

In effect, the being is an energy transforming machine.

What is the purpose of these transformations, locked in the iron circle of the conservation of force?

Why is not everything only electricity or only heat or only light?

Why do these patterns of movement occur?

Why this strange, almost mysterious, analogy between the secular works of nature and those of man, who aspires to improve his machines, to economically transform energy, making them more similar to organisms, those wonderful machines in which power losses are almost negligible?

A sufficiently vulgar comparison can be quite useful in these circumstances:

A ball of soap made of the same type of soap, of the same color throughout, would be homogeneous, as would a universe reduced to light or electricity.

But if the ball of soap is composed of fragments of different color, texture, and size, we now have an idea of the universe as it is: heterogeneous, as Spencer claims.

However, the second ball of soap melted with a mild heat would be homogeneous, as would the Universe if it were composed only of light, heat or electricity.

***

The transformation of the forces must be primarily due to the mobility of ether, to a certain plasticity, to a state of constantly unstable equilibrium, whether electricity or sound, heat or magnetism. That is, instead of seeking the object of the transformations of energy, we need to ask what is that property of the ether due to, this mobility, remotely resembling that of nervous protoplasm that vibrates under the influence of a smaller or intangible corpuscle of a fragrant substance like musk!

Can we speak of a structure of the ether, this medium, whose state of rarifaction seems almost impossible?[64] Evenif we can the same problem remains unexplained: what compresses the ether, who moves this structure, which is the source of electrons?

Religious believers might say: God; a man of science will remain silently meditating.

Summary

In the first book we have attempted to demonstrate the principle that all material phenomena of the organism depend on known physico-chemical forces, and so, we have studied the facts of fundamental unity, explained in the various chapters:

The fundamental unity of macrocosmos, the microcosmos and the mesocosmos.
The unity of the forces of movement.
The unity of matter, the protile, corpuscles, and electrons.
The unity of organisms and protoplasm.
The parallel between the protoplasm and the Cosmos.
Universal thermochemistry.
The Earth considered as a living organisms, in full activity.
The unity of substances in the organic and the inorganic world.
Living beings considered as colloidal minerals.
The parallel between living beings and minerals.
Arquigony, spontaneous generation, Bathybius, Protobathybius, Eozoon, colloidal phosphates, Clay.
The dogma of albuminous substances.
The dominance of mineral substances in biological phenomena.
The parallel between plants and animals.
The fusion of zoology, botany and mineralogy.
The unity of plan in nature.

[64] Spencer accepts this assumption.

	Macrocosmos.	Organic bodies.
	Mesocosmos.	Inorganic bodies.
	Microcosmos.	Living beings and minerals.
Fundamental Unity	Forces.	Plants and animals.
	Matter.	Zoology.
	Protoplasm.	Botany.
	Living beings and stars.	Mineralogy.

Since the review of everything tangible leads to the grandious, but necessary idea, of the fundamental unity prevailing in astronomy, physics, chemistry and the three sister sciences, of the three kingdoms of nature, it would be illogical to accept that organisms are above the known physic-chemical laws and owe their activity, their powers of nutrition and reproduction, to supernatural, mysterious, indeterminate causes.

Instead, each day new procedures are discovered which reduce all the facts, previously called vital, to simple problems of *quantity*, investigating the osmotic constants of the protoplasm, of respiration, nutrition, reproduction…

The life force, which should manifest itself everywhere, with a dazzling evidence, has not been found anywhere, and even if it should be found, it should be a simple variant of heat, electricity or light, a particular form of vibration of the ether, subject to general principles of conservation of energy and the unity and equivalence of forces.

B.

FACTS OF ELEMENTARY OR CELLULAR LIFE.

1. *Physico-chemical properties of protoplasm and the cell.*
2. *Osmotic structure of protoplasm.*
3. *Imitations of protoplasm and cells by inorganic or organic reagents. The protoplasm of clay. Wonderful mirages.*
4. *Mechanical explanation of cell division.*
5. *Imitations of cell division.*
6. *Physico-chemical conditions necessary for the life of lower organisms. Protococos, Trichodesmias, diatoms, bacteria, Infusoria.*
7. SUMMARY.

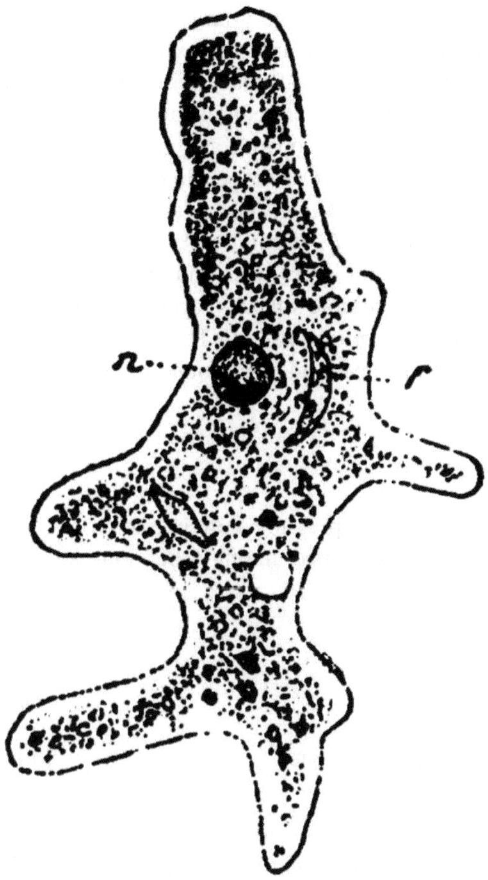

Fig. 14. A blob of protoplasm. An amoeba, a simple organism that constantly changes shape, *n*, nucleus; *f*, food particle (*After Conn*). Greatly amplified. Its general appearance and optical properties are similar to the precipitates that inorganic soluble silicates form in alcohol or ether.

1. PHYSICO-CHEMICAL PROPERTIES OF PROTOPLASM AND THE CELL.

Protoplasm[65] (Fig. 14).

Consistency.—Semi-liquid, viscous, variable, similar to that of glue or water.

Color.—almost always clear. In some fungi is colored.

Refringence.—more refractive than water. Sometimes showing double refraction.

Density.—more dense than water.

Tension.—Like a somewhat thick liquid.

Cohesion. Generally large (80 mmgr. per mmc.) variable. Plasmodial cords are broken under the influence of a traction of 120 to 300 mmgr. per mmc.

[65] According to Rhumbler it has the physical properties of a liquid and is subject to the laws of capillarity. Der Aggregat-zustand und die physikalischen Besonderarbeiten des Lebende Zellinhalts. (Zeitschrift für allgemeine Physiologie. B. I. 1902, pp. 279–388). In these lessons we just give a very brief summary of these profound questions.

Permeability.—Very large for water, some gases and certain dissolved bodies. Impermeable to many dyes when alive. Some live infusoria uptake methyl green. *Reducing power.*—Reduces silver salts and indigo.

Reactions.—Very complicated and confusing.

Diluted acids: opaque and granular.

Concentrated acids: partial dissolution.

Alcohol and heat: opaque and grainy.

Dilute alkali: dissolution[66] or dilation.

Its general appearance and optical properties are similar to the precipitates that inorganic soluble silicates form in alcohol or ether.

Main components.—C, H, O, N, P, S, Ca, F, CI, Si, Na, K, Mg, Fe

Reinke and Rodewald found in the protoplasm of a myxomycete fungus:

Water	71.6
Dry matter	28.4

By means of pressure they extracted 66.7 % of a liquid with a density of 1.209 and that contained among other soluble substances, between 7 to 8 % of soluble albumin.

Dry matter:

Minerals.	29.
Ternary materials.	41.
Nitrogenous materials.	30.

Nitrogenous materials:

Plastin (analogous to casein).
Vitellin (which exists in egg yolk).
Myosin (extracted from the meat.)
Peptone (digested meat.)
Pepsin (digestive ferment.)
Lecithin (phosphorus containing fat).
Guanine.
Sarcin. } Albumin derivatives.
Xanthine.
Ammonium carbonate.

Ternary Materials:

Paracholesterin (extracted from bile).
Special resin.
Yellow colored matter.
Amylodextrin.
Non-reducing sugar.
Fatty acids.
Neutral fats.

[66] Water-insoluble silicates dissolve in alkalis.

Minerals:

Calcium combined with fatty, lactic, acetic, formic, oxalic, phosphoric, carbonic, and sulfulric acids, magnesium and potassium phosphates and sodium chloride.

Conclusion:

Henneguy, Engler, Delage and others conclude that protoplasm is an albuminous substance, but we have already said that such a claim is unsubstantiated, and the basis of protoplasm might be instead an inorganic salt, maybe a calcium phosphate or silicate, which acts as an osmotic and electrolytic apparatus, secreting or absorbing the multitude of secondary bodies that analysis reveals side and that would conform and accumulate between the alveoli or within them, as do the multitude of organic and inorganic impurities in natural clays.[67]

***

Some scholars say that there is no such thing as a single protoplasm, but rather many, but it is likely that they are fundamentally identical in all living beings.[68]

The properties that characterize the physical basis of life, or vital chaos, as Bernard has called it, are essentially:

1st Motility.

2nd Nutrition.

3rd Growth.

4th Reproduction.

Indeed, all protoplasm moves, feeds, grows and reproduces.

There are four necessary conditions for life:

1st Heat.

2nd Humidity.

3rd Oxygen.

4th Reserves.

Same for the protoplasm of human as for that of bacteria.

Before attempting to explain these fundamental properties, we address a very important issue.

2. OSMOTIC STRUCTURE OF PROTOPLASM.

Protoplasm is primarily an osmotic filter composed of countless very small osmotic filters.

Bütschli's studies and those of other scholars have shown that it is composed of small spheres or polygonal alveoli, barely visible with a magnification of 1,000 to 2,000 diameters and operating like osmotic apparatuses with elastic walls, weakly bonded by capillary action.

---

[67] A. L. Herrera. Several publications in las Proceedings of the Alzate Society and the Zollogical Society of France. 1896 to 1903.

[68] Recently it has been said that the protoplasm of plants has fiber very similar to that in the neuroplasm of nerves.

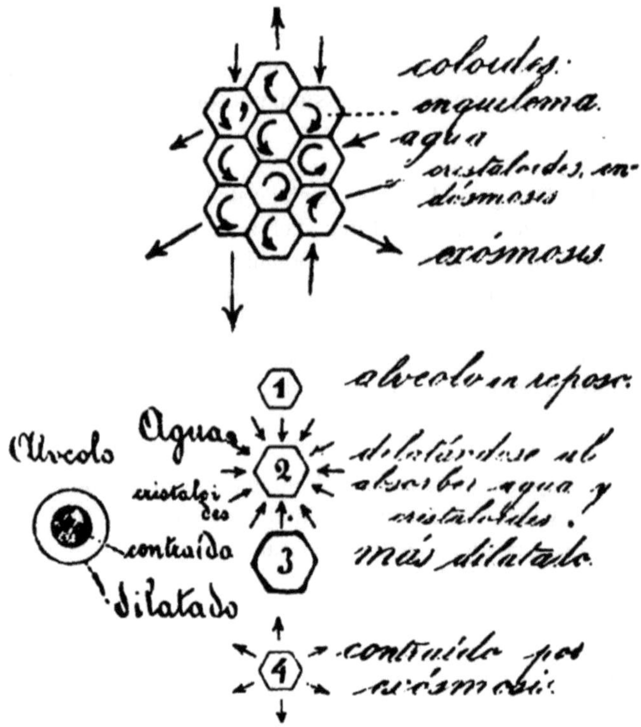

Fig. 15. Schematic figure representing endosmotic and exosmotic currents and contractions and dilations of the alveoli of the protoplasm and sodium oleate emulsions. so that these alternatives are the primary cause of the current and total deformations, and thus of all vital phenomena or movement.

Fig. 16. Optical section of the cortical part of a drop of an emulsion of olive oil and sea salt, showing a very apparent and quite thick alveolar layer (alv), identical to that of certain protoplasm. Magnification: 1,250 diameter (After Bütschli).

They are full of organic acids, salts and other water-attracting bodies, which they attract, through their walls, stretching to increase their content, emptying and deforming,

This is demonstrated very well with alkaline oleate, which shows globules with very elastic concentric walls and which deform, crawl, flutter in water, absorbing it and expelling it (Figs. 15, 16, 17, 18 and 19).

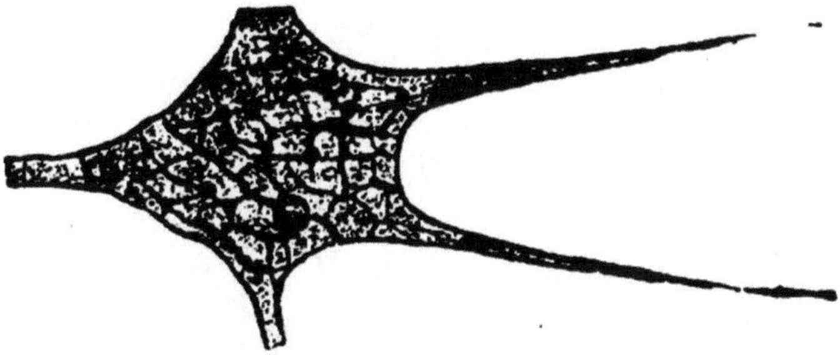

Fig. 17. Palmiform expansion of a living Protozoan pseudopod network. Magnification: 3,000 diameters (After Bütschli). Very similar to certain artificial structures formed from alkaline silicate treated with sulfuric ether.

Fig. 18. Two living protoplasmic strands of a Malva hair. Magnification: 3,000 diameters (After Bütschli).

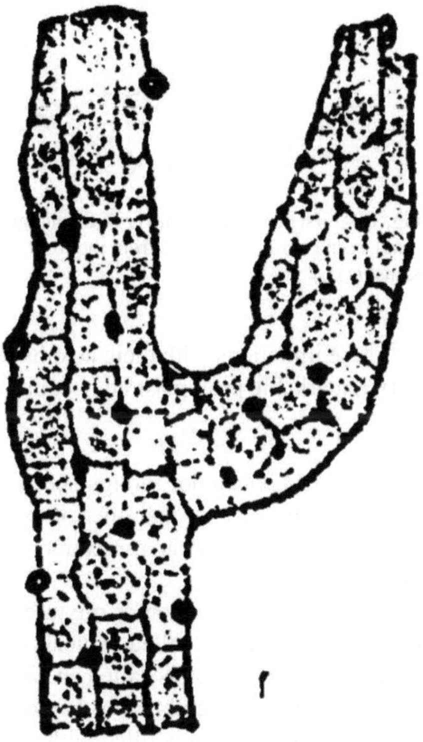

Triturating just a bit of oleic acid to neutral sodium carbonate and water to observe their deformations, is enough to convince oneself that the myriad characters and movements of protoplasm are due to structural reasons, its permeability, its osmotic power, the elasticity of the alveoli, which are arranged in rows, forming fibers, and generally adapt to a multitude of circumstances.

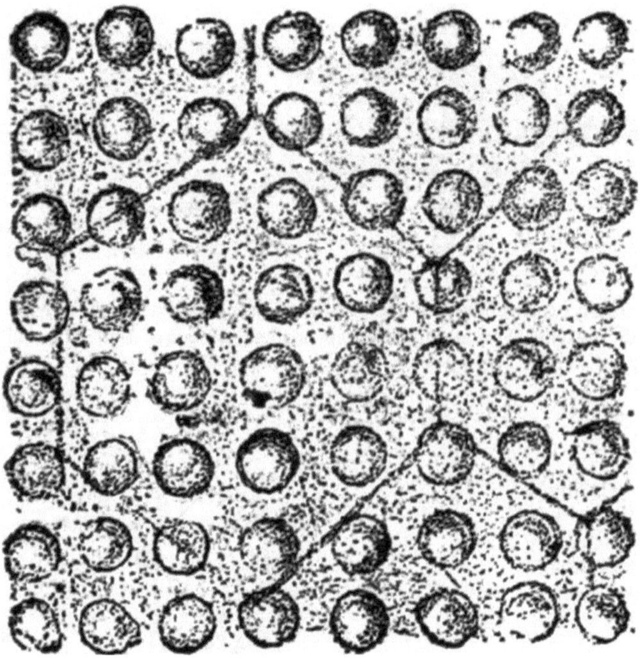

Fig. 19. First sub-cuticular layer of the integument of a flagellated protozoan (*Heteromitus olivaceus*), seen greatly magnified from the front, showing the spherular structure of protoplasm (After Kunstler).

These imitations with oleic acid have shown us over 200 varieties of shapes and very remarkable protoplasmic peculiarities.

***

Neither Mexican nor foreign specialists have been able to find forces or properties in the protoplasm that are not explained by already known physico-chemical laws. Even the synthesis or creation of organic bodies can begin to explained simply.[69]

Osmosis. The diffusion or mixture of two liquids separated by an organic or inorganic membrane is called osmosis (impulsion). The apparatus Dutrochet used to study the phenomenon is called an *endosmometer* (Fig. 20).

It consists of a membranous bag or piece of intestine (A) fixed to the lower end of a tube (a). This bag is filled with a thick solution of gum or sugar or other liquid denser than water, such as milk or albumin, and is immersed in the vessel (B), which contains pure water.

It is soon observed that the water level is rising little by little in the tube, to a height that can reach many decimeters, and descending into the vessel B. This indicates that a portion of the pure water has passed through the membrane to mix with

[69] A. L. Herrera. Le rôle prépondérant des substances minérales. etc. last page.

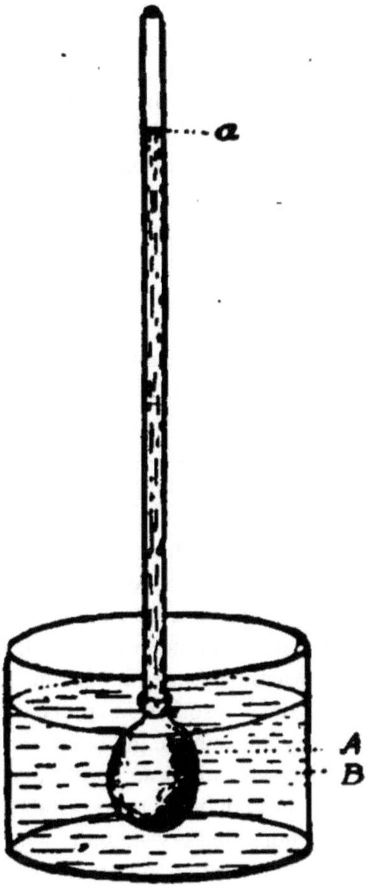

Fig. 20. ENDOSMOMETER. A. Membranous bladder with thick solution of sugar; B. Glass of distilled water. The water enters the bladder, and raises its level in the tube, to the point a. (After Conn). The protoplasm consists of an infinity of extremely small osmotic devices.

the liquid inside.After some time the water in which the endosmometer is immersed contains glue.

A double stream is thus produced and there is endosmosis for the liquid whose volume increases and exosmosis for the liquid whose volume decreases. If you put pure water in the membrane bag and dipped it in glue-containing water, endosmosis occurs from the pure water towards the glue-containing water occurs, which raises the level of the liquid outside.

The height of the rise of the various fluids is variable. Of all the vegetable substances, dissolved sugar is that which, at the same density, has the highest endosmotic power; of all animal substances, albumin does. Gelatin, in contrast, has a weak osmotic power. Endosmotic current is generally directed toward the denser liquid. However, alcohol and ether are exceptions and are conducted from water toward the denser liquids. With acids, depending on whether they are more or less dilute, the endosmosis of water towards the acid or of the acid into the water. These last details show that the passage of fluids through the membrane is probably due, at least in some cases, to a real chemical affinity. And in fact, the exit of water and crystalloid bodies is accelerated (see below) by placing the outer surface of endosmometer in contact with calcined gypsum, which has a high affinity for water.[70]

Dutrochet has shown that endosmotic phenomena occur only under the following conditions: 1st the liquids are heterogeneous and capable of mixing, such as water and alcohol, unlike water and oil; 2nd, The two liquids are of different densities, 3rd, that the separation wall is permeable at leastone of the liquids. All plant and animal membranes are permeable.

Across inorganic walls, endosmosis occurs readily. Pfeffer has shown this by testing copper ferrocyanide, calcium phosphate and Berlin blue,[71] although the latter do not stand up to an osmotic pressure, which calcium silicate may support better.

And strangely enough,"organic membranes are disrupted quickly when endosmosis ceases." This is what happens with alkaline oleate emulsions: these could not serve as the protoplasm's osmotic devices, because they would disrupt easily under the influence of acids, alkalis, and enzymes. The same applies if the protoplasm was formed, as assumed by the authors, from an albuminous substance resembling egg white.

Endosmotica absorption takes place in plants in the root hairs. Lower animals are nourished by endosmosis. In higher animals there is true absorption. For example, the coloring substance called ruby, when taken inside, penetrates the bones and tints them red. If a liquid is in contact with a cutaneous surface stripped of its epidermis, or with a mucous membrane, the liquid passes into the blood vessels by endosmosis. Fats are not absorbed in appreciable quantities because membranes do not get wet, but they do so in the gut once emulsified by the digestive juices.

Dialysis. This is an application to chemical analysis of the uneven diffusion of fluids through membranes. Graham, who discovered it in 1861, divided bodies into two groups, the crystalloids, which have the property of crystallizing (salts), and the colloids or the non-crystallizablesubstances (gum, starch, dextrin and albumin, which crystallize only under very special conditions).[72]

Crystalloid solutions are not viscous, but always sapid and spread easily through membranes and porous walls: colloids, in contrast, are viscous, tasteless and have little tendency to spread.

Graham's diffusion apparatus (Fig. 21) is a modified endosmometer,

called a dialyzer. It consists of a glass or gutta percha drum 25 inches in diameter, sealed at its base by a membrane, thus forming a sort of sieve. In this is deposited a mix of colloidal and crystalloid substances (egg albumin),[73] and the latter passes through the membrane and enter the water in the outer vessel that the screen floats on. The force that causes the body to pass through the membranes is called osmotic pressure, and varies according to many circumstances. It may be, as mentioned, a

[70] A. L. Herrera. Diálysis química. Aplicaciones del sulfato de cal. Thesis for the bar exam. Mexico. 1889.

[71] Jones. The Elements of Physical-Chemistry, p. 187. Prussian blue is also called Berlin blue, and is prepared with potassium ferrocyanide and iron sesquioxide.

[72] For Wroblewsky's procedure, using evaporation through a membrane.

[73] Contains many dissolved salts.

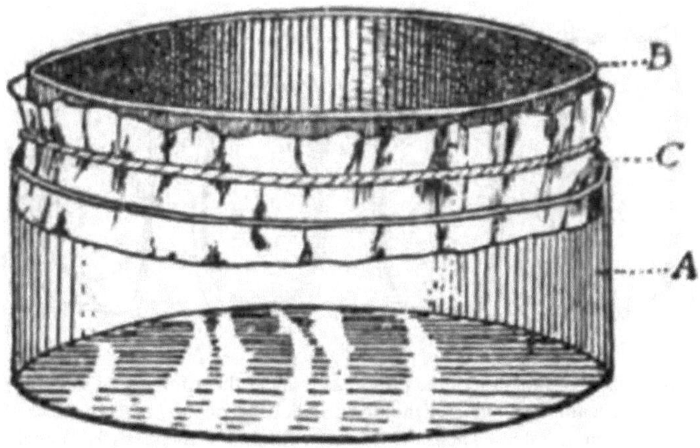

Fig. 21. Dialyzer – A, pure water. B, mixed solution. C, membrane separating both liquids. (After Conn.).

chemical affinity. Notwithstanding, it was found that the osmotic pressure of a solution of cane sugar is exactly equal to the pressure of a gas having the same number of parts in a given volume, at the same temperature. Under the same conditions, a dissolved particle exerts the same osmotic pressure as a particle of gas. It is thus believed that the particles of liquids and dissolved substances are moving in all directions, like gas molecules, according to the kinetic theory. This could be likened to the motion of a swarm of bees enclosed in wire mesh box.

Dissociation of electrolytes. We have seen that dialysis separates colloidal and crystalloid bodies, but there is another important cause of dissociation: acids, bases and salts, when dissolved in water, dissociate into ions. For example:

$$HCl = H^{+} + Cl^{-}$$

Hydrochloric acid, is split into hydrogen (+) and chlorine (−).

$$KOH = K^{+} + OH^{-}$$

Potash, is divided into potassium (+) and the OH^- radical.

$$KCl = K^{+} + Cl^{-}$$

Potassium chloride, is divided into potassium (K^+) and chlorine (−).

The positively charged bodies are called *cations* and negative one *anions.*

It has been noted that many reactions are impossible without this separation: thoroughly dried hydrochloric acid does not decompose carbonates, dry chlorine does not combine with sodium, etc.

Thus, electrolytically dissociated bodies have maximum activity and react strongly in the protoplasm, which is a mixed osmotic and electrolytic apparatus.

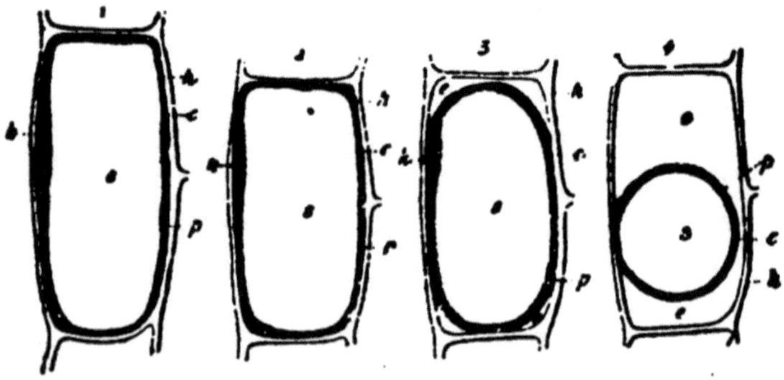

Fig. 22. Plasmolysis or contraction of the protoplasm under the influence of salts. 1. Young, half-grown cell of the cortical parenchyma of the floral stalk of *Cephalaria leucantha*.-2. The same in a solution of 4% sodium nitrate.-3 The same in a 6% solution.-4. Idem at 10%, *h*, cell membrane, *p*, primordial utricule, *k*, nucleus of the cell, *c*, clorophyll body; *s*, cellular sap, *e*, saline solution. (After *Hertwig*.) Artificial plasmodia from colloidal silicates also contract under the influence of concentrated salt solutions.

Loeb and others have observed that the monatomic ions Na^+, Li^+, Rb^+, Cs^+, are very favorable and necessary for life, and the di- and triatomic ones less so, or even poisonous; making it necessary that they be present in certain relations to sustain life, as occur the seawater.

According to Hardy the diatomic and triatomic ions dissolve colloidal substances in living beings.[74] However, salt water containing the monatomic cation Na^+ dissolves many nucleins and albumins and is a powerful resource, applied by injection, in cases of heavy bleeding, anemia, and general loss of strength and blood or a part of the serum of the same.

This action of salt may be simply due to the removal or addition of water to the protoplasm (Fig. 22), and so much so that the same saline solution is poisonous or benign, depending on their degree of concentration, and attract or repel motile cells[75] (for example, sperm).

***

Movements of the protoplasm. Require the presence of oxygen, which produces a very favorable heat release for osmosis, varying the viscosity of liquids in various parts of the protoplasm, which determine the very active diffusion currents.

Every protoplasmic alveolus is filled with a liquid called enchylema, which is very rich in proteins, organic acids and salts, and which has a great affinity for water. They absorb this, and the alveoli, whose colloidal walls are elastic, expand, taking on a spheroidal shape which results in a movement of expansion, like a rope that

[74] 1 Hoeber. Neue Forschungen über die Beudeutung der Neutralsalze für die Funktionsfahigkeit der thierischen Protoplasten. Biochemisches Centralblatt. Bd I, No. 13, p. 497.

[75] Livingston. The Role of Diffusion and Osmotic Pressure in Plants. 1902, p. 148.

stretches when wet. But then the intra-alveolar pressure is increased and exosmosis predominates over endosmosis (which also fluidizes the enchylema by absorbing water), and a movement of contraction is observed. The same series of phenomena occurs constantly, especially under the influence of oxygen and thus all movements of protoplasm result, its displacement in the direction most favorable for osmosis, the vibration of the vibratile extensions, which are very flexible osmotic pipes, etc. This full set of phenomena is observed very well in alkaline oleates, which sometimes form very thin, floating and contractile films. The pressure produced by the breath of the observer causes them to palpitate rapidly. They also tend to form irregular vibrating lashes which oscillate with a sort of rhythm, with a period of 5 to 10 minutes.

Of course these movements are not due to the supposed life force.

Some authors relate them to surface tension and electrical phenomena, but in any case they are mainly due to the osmotic structure of protoplasm.

Nutrition of the protoplasm. This is accomplished by endosmotic absorption, after secretion of gastric juices or enzymes that dissolve albumin or starch, and render them diffusible. Salts and organic bodies that penetrate the alveoli undergo a multitude of transformation and from them the elaboration of countless new compounds results. All this is still very unclear. Probably the alveolar complex is a porous body, like a platinum sponge, and when the absorbed elements come into intimate contact inside the tiny alveoli (which are only visible with a magnification of 1,000 to 2,000 diameters) kinetic energy is condensed forming new molecules which are trapped there or are liberated after the rupture of the osmotic networks. Colloidal inorganic ferments show something similar, forexample, colloidal platinum metal, due to its fine particle size works like ferments.

The dissociation of ions should influence that some atoms move more than others within the osmotic filter.

Growth of protoplasm.

The only theory that explains it is very risky and assumes that some inorganic acid (silicic) is combined with an inorganic base salt forming a colloid impregnated with inorganic and organic substances, which are absorbed or secreted under specific conditions.

Speaking of clay protoplasm can tell you how to form a precipitate of silicon or an inorganic silicate, which *grows* a little, absorbing *increasing* amounts of acid and base.

***

The protoplasm of plants grows using inorganic elements, air, water and salts and is in summary the result of a chemical reaction, which can be induced by adding drops of sulfuric acid and barium nitrate solution in a watch glass. The precipitate of barium sulfate will increase as long as the addition of reagents is continued.

However, this phenomenon is as natural as the growth of a plant that lives using inorganic substances. The mechanism of formation of the colloidal precipitate that

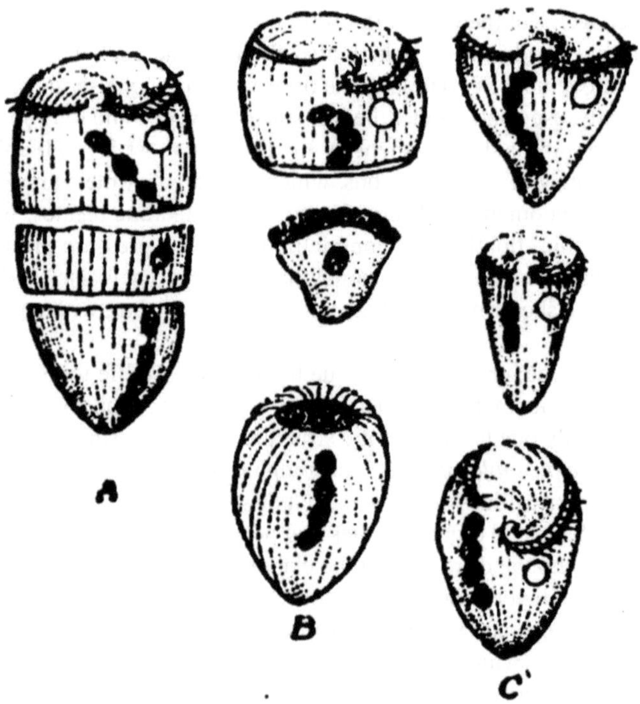

Fig. 23. Infusoria divided into three fragments, each of them contains a partial nucleus. All continue living and soon take the form primordial as shown in c. (After Conn).

forms the basis of protoplasm is still not known with certainty, but all presumptions are in favor of a natural origin.

Reproduction. The protoplasm and the cell reproduce under favorable conditions of nutrition and always having a nucleus or a nuclear fragment.

If an infusorium is divided, the fragments devoid of this organ soon die, and those which retain a small part of the nucleus continue to live (Figs. 23 and 24), as if nuclein were essential for life, perhaps due to the phosphoric acid they contain in small amounts. Various reasons have led them to be considered as reserves, but there is still not sufficient evidence for this theory. However, Van Beneden and Brass have observed that the essential parts of the nucleus are dissolved when the cell is starving.

We shall not describe in detail the phenomena of cell division, which is already studied by students in courses in physiology or natural history, and merely copy a few very clear figures from "Notions of Bioiogy" by Conn, and noting that studies on composition of the nucleus and its figurative elements are still very imperfect and based primarily on coloration, there being no two authors who agree as to how to interpret them (Figs. 25 to 34).

According to Houssay protoplasm and the nucleus have an alveolar structure and are not separated by a true membrane. The agglomeration of the nuclear mass is

Fig. 24. Cell divided into three parts, containing a nucleus only in number 2. This fragment soon acquires the primitive form and continues to live. The other two fragments live for a time and then die without reproducing. (After Conn.).

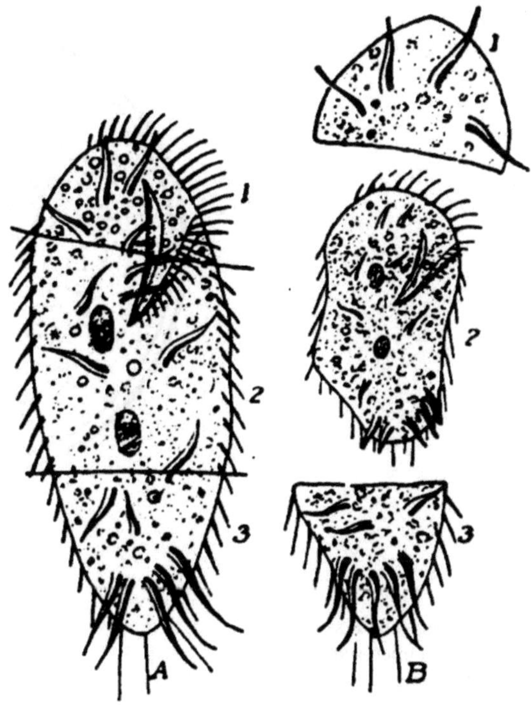

Fig. 25. Ovule. Showing the cellular substance and the nucleus, containing a great number of chromosomes and a nucleolus. (After Conn.)

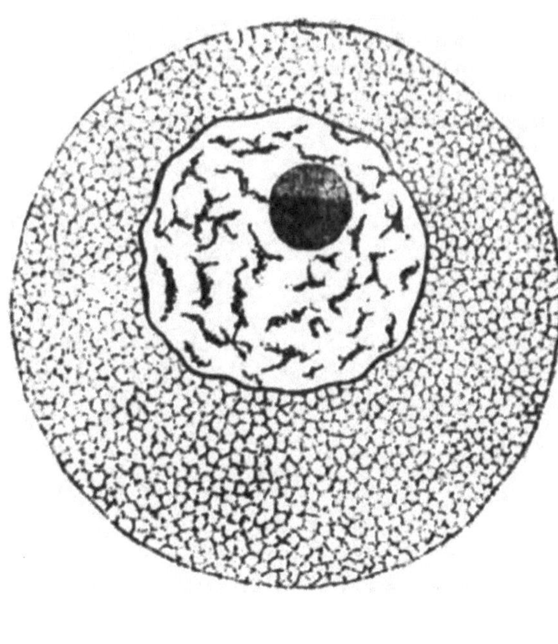

Fig. 26. A cell. *cw*, cell wall, *pr*, cell substance, *n*, nucleus. (After Conn.)

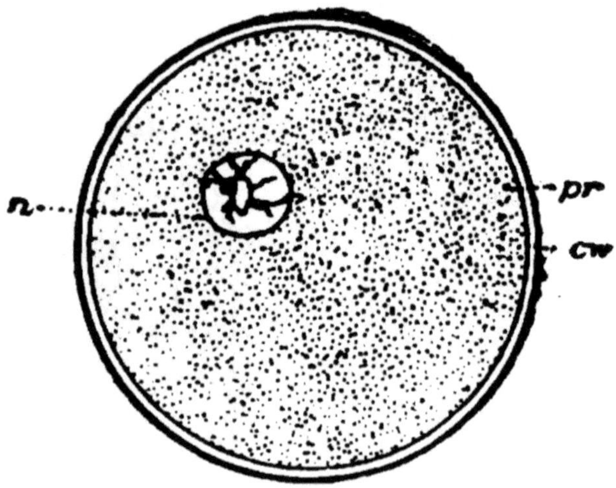

Fig. 27. The cell, as seen with modern microscope. *a*, protoplasmic network, *b*, liquid, contained in the nets, *c*, nuclear membrane, *d*, nuclear network, *e*, network of chromatin, *f*, nucleolus; *h*, centrosphere, *i*, void or vacuole, *j*, inert bodies. (After Conn.) All of these structures are formed from even finer materials, not shown in the drawing.

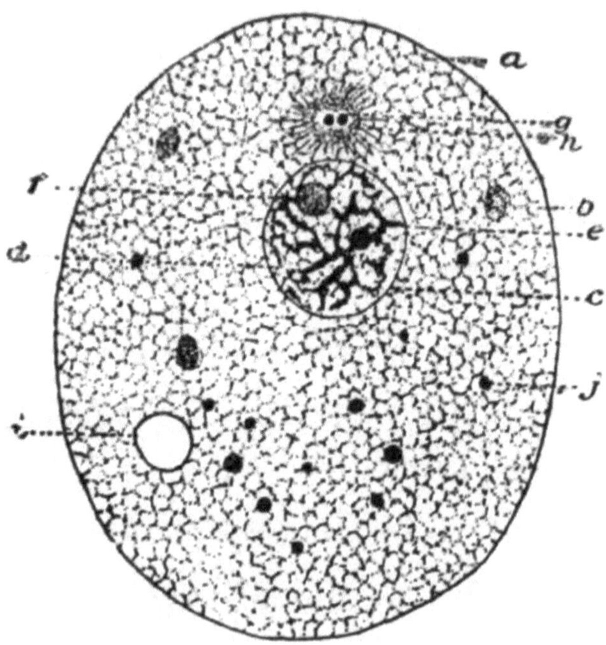

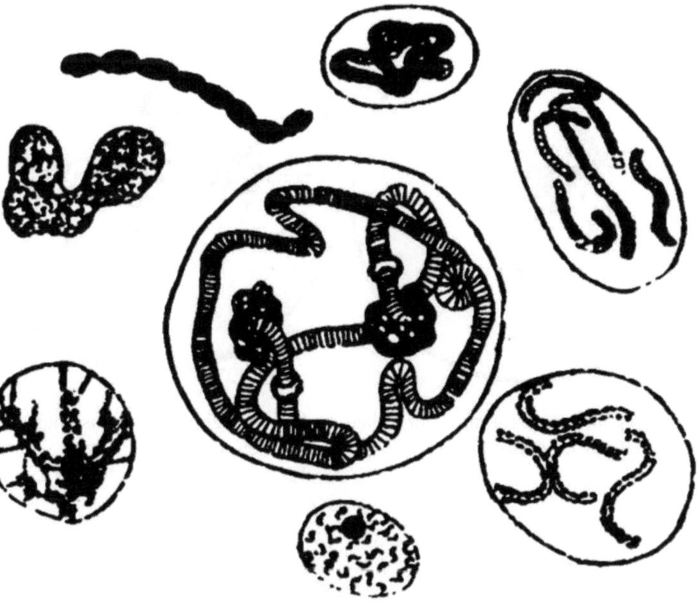

Fig. 28. *Different forms of nuclei.* The chromatin has various aspects. (After Conn). In the central figure a nuclear cord or espirema identical to artificial structures formed from calcium silicate.

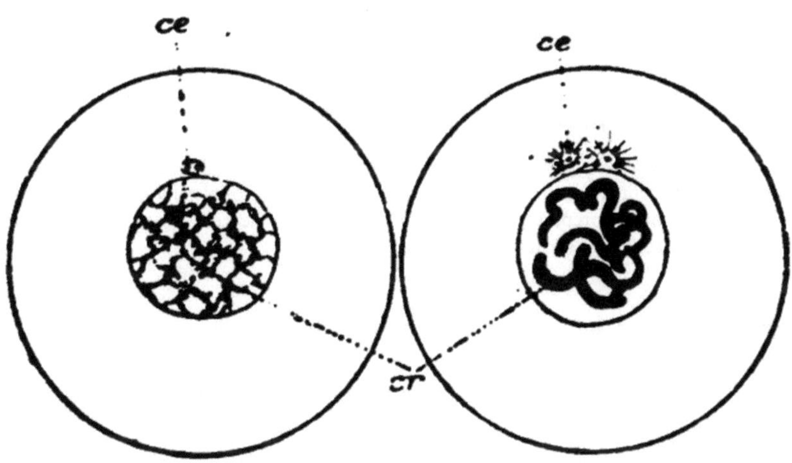

Fig. 29. Cellular division, mitosis or karyogenesis. Left: resting period, the chromatin *cr* affect the shape of mesh inside the nuclear membrane, two directing spheres with centrosomes, *ce*. Right: the chromatin rotates forming threads or chromosomes, *cr.* The centrosome, *cc*, shows radiating fibers very similar to those of certain artificial silicate colloids.

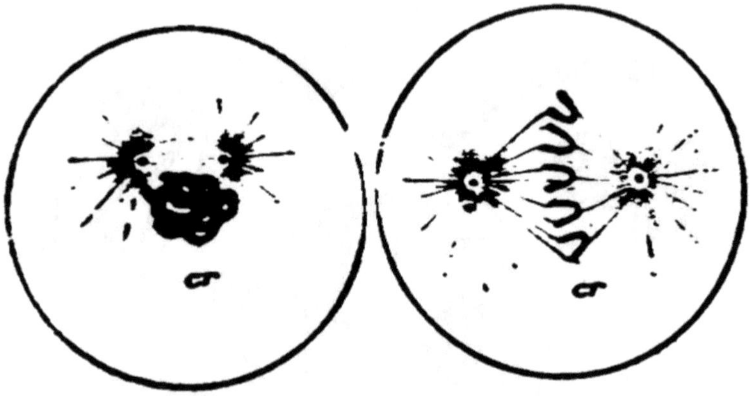

Fig. 30. Left: separate centrosomes, but linked by fibers. Right: the centrosomes are separated, and the equatorial plane of chromosome, *cr*, is between them (After Conn).

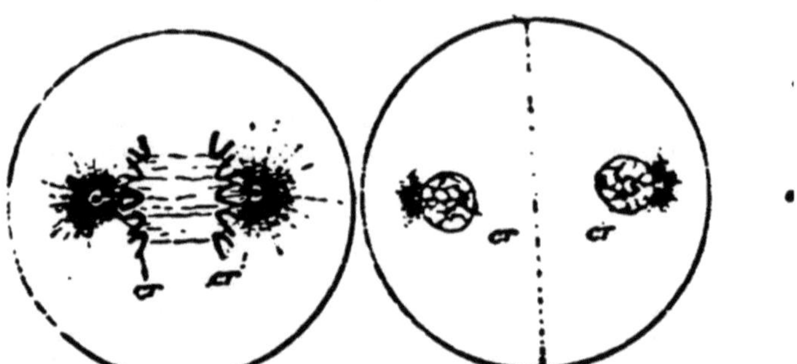

Fig. 31. Left: the two halves of the chromosome separated from each other. Right: final period. Two new nuclei in which the cromosomes have taken the form of a network. The centrosomes have divided previously, preparing for the next division and the cell begins to divide itself by means of a wall of cellulose. (After Conn).

the result of chemical and physical qualities and because these substances are more compact in the center of the cell.

It has been sought to make cell division a phenomenon subject to certain unexplained fatality and inexplicable perseverance, but according to

Houssay, direct and indirect division depends on the circumstances and this can be seen in the cells of the same animal; "None of the acts of indirect division is exempt from variations, they should not be considered as necessarily given, unchanging and simple, from which we can deduce a general explanation with respect to the multiple properties of animals. This is a natural phenomenon; it can be reduced to the laws of physics and should not serve as an initial basis for speculation."[76]

[76] Houssay. La Forme et la vie, p. 538.

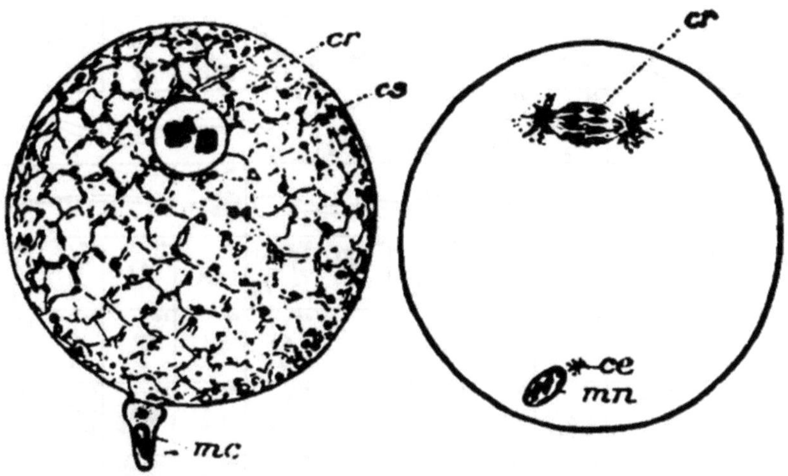

Fig. 32. Fertilization of the egg. *Cr*, chromosome, *cs*, cellular substance, *mc*, masculine reproductive cell after entering the egg. Right: The centrosome of the ovum divided, the masculine cell having entered the egg with its centrosome (After Conn).

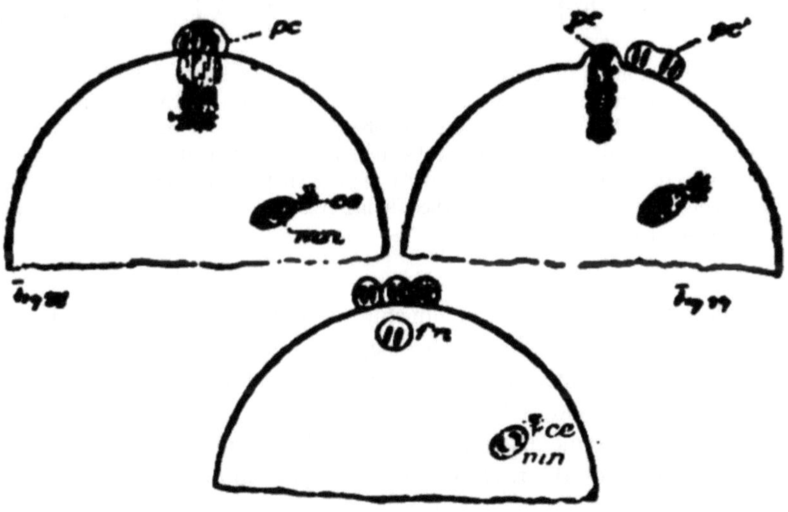

Fig. 33. Details of the fertilization and division of the egg: *pc*, first polar cell, *pc"*, second polar cell, *rn*, the two chromosomes retained in the egg, forming the female pronucleus (According Conn).

For us, cell division consists simply in that the tapes or cords of nuclein, or reserves, fall under the influence of the traction exerted by the protoplasm of the cells, to absorb water or certain salts, and as osmotic conditions change.

Although we say that this hypothesis has not been proven yet, we confess that the phenomenon of cell division in our eyes has lost its great prestige, since we can

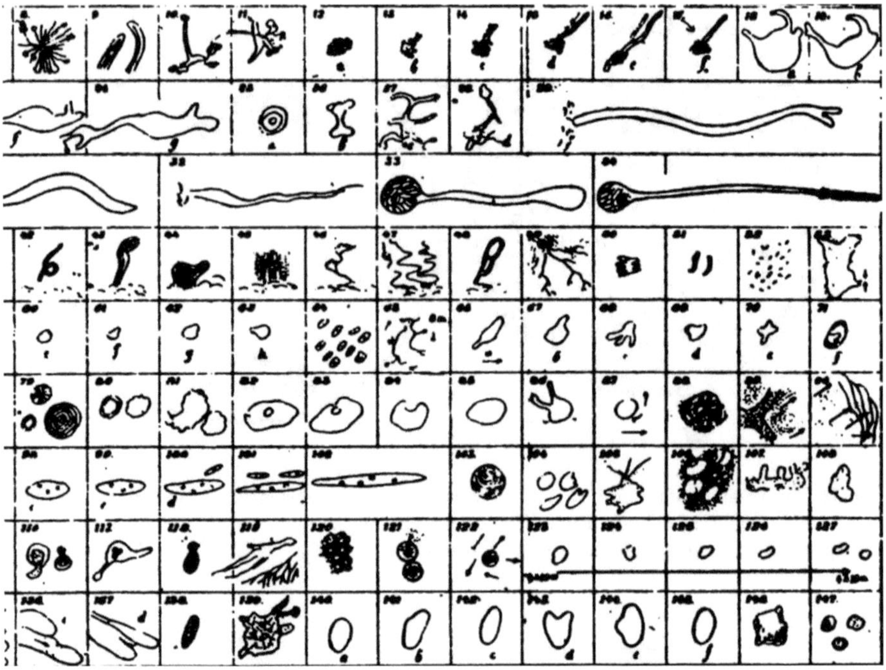

Fig. 34. Imitation of many structures, movements and variations of the protoplasm and microscopic organisms, by means of alkalis and oleic acid. The arrows indicate the direction of the currents or movements.Note the imitation of microbes (numbers 45, 51, 52, and 93), or vibrating lashes (numbers 53, 55, 65), of moving amoebas (numbers 181–184), of plasmodia (number 90), and cells with interior filaments (number 152). All of these are formed instantly while the observer watches or evolve in very alkaline water, where living organisms cannot live and are not present. Oleic acid contains inorganic impurities.

imitate with sodium silicate and calcium chloride, precipitating under certain conditions, and also with wax solution in ether, deposited in water. It is a simple matter of consistency, strength, center of resistance, and does not deserve many descriptions and technicalities that confuse the pupil's mind.

If you employ the same detail to describe the formation of colored precipitates, of crystallization, of colorful reactions of bilirubin and nitric acid, chemical science becomes intractable.

Other difficulties are of concern to the investigator of natural phenomena, so much so that after giving paramount importance to the nucleus, one begins to understand its uselessness without the protoplasm.

In fact, a cell without a nucleus or without protoplasm, is unable to reproduce perfectly.

"The nucleus needs protoplasm, and protoplasm the nucleus." Life result from this asociation.[77] The nucleus is more condensed plasma, and not a separate special entity, but exists in almost all beings, if not in all of them. Even in artificial cells it has been seen with greater clarity, and no one would say that these have extraordinary qualities or composition.

In our opinion, the plasma and the cell should be studied simultaneously, as all hydrostatic phenomena are studied in hydrostatics and all light in a chapter on optics.

In short, it seems inexplicable that great attention is paid to the indirect division and not to the direct division of the fragments of the nuclear tape, which no doubt are being consumed and regenerated during the life of the cells and serve as something, not only as bystanders....

Delage has said that the most interesting thing about direct division is the direct division of chromatin and cytoplasmic fragments, as we have already noted, according to Van Beneden and Brass, the chromatin can be dissolved in some cases, during starvation, decreasing the nucleus volume up 44.6 percent,[78] which provides proof of its nutritional qualities, its reserve role, serving as food for various bacteria and other parasites of the nucleus, called nucleophages. Some of these fully absorb the chromatic substance.The nucleolus, according to Strasburger and Guignard, dissolves in the nuclear sap, like starch in plant juices, and is absorbed by the chromosomes. In addition, the chromatin appears to be most abundant during cell division and is believed to partly diffuse into the cytoplasm.[79]

***

3. IMITATIONS OF THE PROTOPLASM AND THE CELL BY MEANS OF ORGANIC OR INORGANIC REAGENTS. THE PROTOPLASM OF CLAY. WONDERFUL MIRROR.

4. MECHANICAL EXPLANATION OF CELL DIVISION.

5. IMITATIONS OF CELL DIVISION.

A new science, PLASMOGENY made its appearance in the late nineteenth century. It aspires to mimic living organic structures using reagents.

[77] Calkins. The Protozoa, p. 303.

[78] See Loukianow's studies. L'inanition du noyau. Revue Scientifique, 23 October 1897, p. 514.

[79] Henneguy. La cellule, pp. 363 and 364.

The main imitations obtained to date are as stated below:

Years.	*Authors.*	*Reagents.*	*Results.*
1824.	Dutrochet.	Albumin subjected to the action of an electrical current.	Globules (by coagulation?)
1840.	Ascherson.	Albumin and fats.	Globules (by the formation of soaps, upon the mixing of alkalis with albumin and fat).
1864 } 1867 }	Traube.	Gelatin hydrated at 150 °C and dissolved tin.	Cells of gelatin tannate operating as osmotic apparatuses.
1868.	Rainey.	Rubber and zinc chloride.	Vacuolated cells.
1872.	Harting.	Calcareous precipitates in colloidal media.	Concentric and radiated structures, pseudo-shells. Otoliths.
1884.	Quincke and Bütschli.	Soaps during their formation; rancid oil triturated with salt or sugar (Fig. 16); jilol and soap, water and oil, gelatin.	Very important: amoeboid movements, alveolar structures osmotic currents, vacuoles, orientation of alveoli.—These experiments have served as the basis of *plasmogeny*.
1901.	Leduc.	Potassium ferrocyanide and copper sulfate, or potassium ferrocyanide and gelatin.[1]	Complete cells with nucleus and nucleolus, cytoplasm and membrane. Sensitive to a multitude of reactants Osmotic currents which are paralyzed by drying.
1889 to 1903.	Herrera.	Various organic or inorganic reagents. Tannin powdered over gelatin. Oleic acid and alkalis or alkaline carbonates. Oleic acid floated on lime water. Oleic acid dissolved in carbon sulfur in the bottom of a glass with ammonia water. Metaphosphoric acid and calcium chloride in salt water. Soluble phosphates and soluble calcium salts.	Cells, spermatozoa, protozoa. All immotile and invariable. Hundreds of forms, structures, movements, evolutions very similar to those of natural protoplasm (see Fig. 34) Divisions, vibrating movements, nucleated cells, with inferior filaments, palpitating films, tubes and contractile filaments. Amoeboid movements visible to the naked eye. Very clear amoeboid movements, contractile vacuoles, conjugation and direct division. Plasmodia, amoeboid movements, yeasts, polygonal alveoli.

[1] As ferrocyanide is prepared by calcining common potassium carbonate and organic bodies, hair or leather, it almost certainly contains some amount of soluble silicates. Copper silicate produces the same cells of Leduc.

Comparison and reflections

A multitude of insoluble viscous bodies closely resembling protoplasm and varying in the details of their structure and movement according to their mode of preparation and many other accidental circumstances (impurities) are formed.

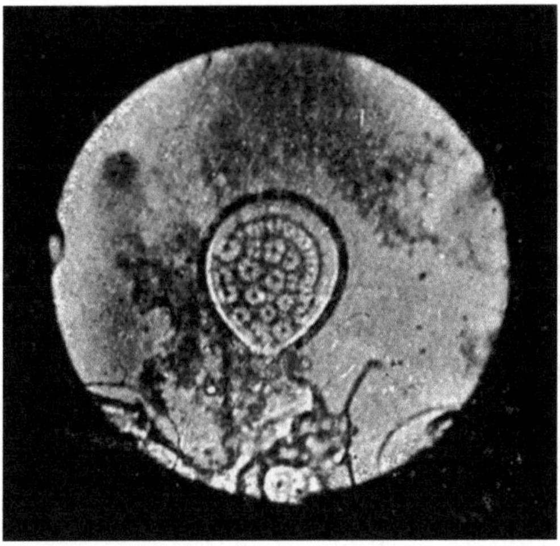

Fig. 35. Imitation of segmentation from oleic acid and neutral sodium carbonate. Upon absorbing water the alkaline oleate particles expand more the lower the resistance which opposes their expansion.

It is all very easy to reach exaggerated conclusions. Bütsclili attributed to forming soaps the structure of protoplasm, and we, to the alkaline oleates, which really have a striking analogy with natural emulsions. Later it seemed more logical to us to look at the inorganic structures of calcium phosphates and colloidal silicates.

In any case, *the physical properties of protoplasm have been faithfully imitated with organic or inorganic reagents and PLASMOGENY is already an experimental science.*

As metaphysics does not easily abandon its stubborn preoccupations it has been said that artificial structures do not show the phenomena of cell division. Actually, this has been emulated by Gallardo, by Rhumbler (alveoli of grenetin) and Bütschli (the same). We accidentally discovered a very simple method: drops of ethereal white wax poured, are added to pure water, forming radiations and other macroscopic structures very similar to the mitotic ones or those of indirect division, which are due to, according to Rhumbler, the tension exerted by the rows of alveoli of the protoplasm, acting on the nucleus of the cells.

Sodium silicate treated with salt water forms nuclei with granular radiations.

***

The study of these imitations proves Berthold's theory according to which the cell is composed of concentric spheres, as displayed in the oleates, which dilate more or less, according to the resistance of the limiting membrane, making the contents in the inner layers thicker and thicker. If a sodium oleate particle prepared with carbonate dilates, absorbing water, a kind of egg in the process of segmentation is formed (Fig. 35), the interior particles dilate less as they are in contact with a thicker medium upon crossing the outer membranes, and due to the lack of space to expand and the corresponding pressure. This is why the nuclei do or do not take the shape of the cell, according to the conditions of pressure and osmosis, and even deform under the influence of salts which dissolve the limiting membranes.

CLAY PROTOPLASM. – MARVELOUS ILLUSIONS.[80]

[*a.*] *The presence of silicic acid in the reagents.*

The observed irregularities upon preparing certain imitations of protoplasm force us to recognize and study our reagents, parlicularly metaphosphoric acid, tannin, oleic acid, egg white, and as many times we have discovered traces of silicic acid, proceeding from sample preparation, or from the long stay of the substances in cheap glass jars, or, the extensive presence of silica in the nature, it thus occurred to us to examine under the microscope the various precipitates of silica and insoluble silicates.

The result was very interesting, but first we should take note of the following explanations:

[*b.*] *Silicic acid exists in protoplasm.*[81]

Henneguy explicitly says this.

According to Carpenter,[82] silicic acid penetrates to an extraordinary degree, the whole structure of the *Equisetaceae* (horse tails), even after destroying the organic matter with nitric acid, a persistent skeleton remains. In some species this mineral forms 13 percent of the solid matter and 50 percent of the ashes.

The same Mr. Carpenter says that silica permeates the epidermis of many plants and form, in residue after incineration, a skeleton of the cells of the cuticle, hairs, stomata, etc., as well as the thorns of the grasses, especially rice, in blades of grass having the appearance of rows of little cups.

The epidermis and scaly hair *Deutzia scabra* also contain a large amount of silica.

The spicules of sponges have a cavity filled initially with organic matter and "seem to have been originally a tissue-filled segment, whose form is determined by the spicule."[83]

Finally, the algae called Diatoms have an envelope of silicic acid.

Where silica and silicates have been found.	Authors.
Bark [Caprinus, Acer]	Löw. "Nutrient Minerals."
Flax. 28 percent of the ash.	" " "
Hair and feathers, forming an organic compound.	" " "
Animal tissues, greater in young tissues.	" " "
Pancreas. 12 percent of the ash.	" " "
Human hair: 0.10 to 0.02.	" " "
Plant leaves.	" " "
Grain straw.	Otto. "Organisches Chemie."

[80] The details of this part are not mandatory for students, up to number 6. (Physico-chemical conditions necessary for the life of inferior organisms.).

[81] La Cellule, p. 25.

[82] The Microscope, p. 418.

[83] Ibid, p. 607.

Where silica and silicates have been found.	Authors.
Cane bark.	” ” ”
Juncos. Bamboos.	” ” ”
Potato.	” ” ”
Blood.	” ” ”
Egg white, in large numbers	” ” ”
Microscopic organisms.	” ” ”
Gravid uterus.	” ” ”
Guano.	” ” ”
Coal.	” ” ”
Silicified wood or jilópalo.	Fliche. Bull. Soc. Sci. Nancy. III. 4.
China guts.	Stein.
Quercus imbricaria bark. 0.30 percent of the ash.	Ekert.
Diatoms: contain iron silicate and silicon.	Frankland and Smith.
Medullary rays.	Bargagli.

[*c.*] *Silicic acid exists everywhere. Its polymorphism.*

In the depths of the sea, at 8,000 meters,[84] in limestone terrain, in alkaline springs[85], in lavas, etc., etc. The great work of Dana [*A system of Mineralogy*] has 1104 pages, and the study of silicates occupies 409, describing over 300 varieties.

According to Joly and Curie,[86] after oxygen, silicon is the most common body in the known part of the globe, in all terrains, and forms the basis of the earth's crust. It is very common in clays with a soapy (Moroccan soap[87]) or gelatinous consistency (Plombierite[88]), which hardens little by little. Mr. Bárcena mentioned an example of liquid opal, found in a mine in Guanajuato. Four or five months later several nuclei formed in which the opaline sheen became visible.[89]

Other strange forms:

Agate, with colored nuclei and concentric layers of distinct osmotic power.

Dendritic agate and moss-like agate, with branching tree-like structures.

Granular quartz. Cellular structure.

Beekite. Takes the form of calcareous shells.

Opal. Reniform, stalactitic, tuberous, like protoplasm, usually doubly refringent.

Fiorite. Sometimes fibrous or stringy, deposited in part by the action of plants.

Miguelite. Capillary or filiform.

Geyserite. Like a sponge or vegetation

Saponite. Like soap.

---

[84] Robin. La Terre, p. 71.

[85] Robin. La Terre, p. 111.

[86] Ency. chimique de Frémy. T. II, p. 113.

[87] Dana. A system of Mineralogy, p. 681.

[88] Dana. A system of Mineralogy, p. 670.

[89] Bárcena. Geology, p. 79.

[*d.*] *Imperfections of analysis.*

It is very difficult to recognize very dilute colloidal silica and silicates, because their reactions are highly reduced and are very uncertain in many circumstances. The gelatinous precipitate produced by other acids do not occur when the mixture is well stirred and the reagents highly concentrated. When silicates are calcined, or anhydrous silica escapes, as a very faint vapor, or soluble salts are formed, if there alkalis present, as in the ashes of organisms, and the body in question goes unnoticed. To form hydrofluorosilicic acid a large enough dose of silica and extreme care in manipulation is needed.[90]

Moreover, in chemistry laboratories there are many glass and porcelain objects, which have a silicate base, which very frequently contaminate reagents[91] and water. And this is not all, in the colloidal salts of silicic acid, such as silicargol, classical reactions are hidden and the recognition of the components is almost impossible. Even if they were not, the chemical analysis of tissues are still very imperfect, and, for example, arsenic, found only in a few organs, has been discovered in almost all of them, through the use of very thorough new methods. Because silicon exists throughout nature it would be strange if all organism did not contain it, in their various parts, bathed in water, blood or sap, which are never exempt from the ubiquitous metalloid.

[*e.*] *Importance of silicates for nutrition.*

Raulin's nutritional liquid, widely used in bacteriology laboratories for the cultivation of lower organisms contains 0.07 gr potassium silicate. The addition of this to the solutions in which *Aspergillus* is grown increases the dry weight of cultivates in the ratio of 1.2 or 1.4 to 1. (Raulin.)

There is no doubt that all laboratory culture media contain silica, from glass vessels and various impurities. We have already mentioned that phosphoric acid is not free of silicates.

[*f.*] *Silica and inorganic colloidal silicates form large deposits in the sea and on earth, and their physical appearance is very similar to protoplasm, as are some of their properties of assimilation.*

Clay (hydrated aluminum silicate) has a more or less plastic consistency and abounds in nature.Colloidal silica exists in the geysers at the bottom of Berkshire and Wilton Chalk, England and elsewhere, as in the inferiol asli. of siliceous sponges and other lower organisms, probably combined with organic materials.Marine deposits, according to J. Thoulet, which represent a much more pure clay the greater the depth.[92]

[90] Ribot. Docimasie. 1, p. 579.

[91] Imitations of protoplasm prepared with alkalis and oleic acid, albumin, tannin, gelatin, metaphosphoric acid, almost always contain traces of silicates. The same is true of copper ferrocyanide.

[92] C. R. Acad. Sci Paris, 3–10 Dec. 1894.

ASSIMILATION PROPERTIES.

Species of silcate or silicon.	Bodies which absorb or with which the mixture is combined.	Authors.
Colloidal silica.	Silver	Hanriot.
Clay.	Various bodies which no type of leaching will separate.	G. Vogt.[a]
"	Water and ammonia.	Liebig[b].
"	Calcium, in aqueous mixture.	–
"	Ammonia.	Pichard.
"	Fats. Soaps.	Various.
Colloidal silica.	Alcohol.	Joly and Curie.
" and alcohol.	Ether.	"
" "	Benzene.	"
" "	Carbon disulfide.	"
" " etherea	Oils.	"
" "	Glycerin.	"
" colloidal.	Sulfuric acid.	"
" "	Nitric ".	"
" "	Formic ".	"
" "	Acetic ".	"
" "	Alkalis, forming colisilicates.	"
" "	Gelatin.	"
" "	Albumin.	"
" "	Casein.	"
" "	Coloring materials.	"
Clay	Multitude of salts and metals. (200 varieties known as silicates. According to J. Thoulet, clay is the last form of individual existence of all minerals.[c]	Various.
"	Iron, oxide. Forms various mixtures.	
" smectite	Oil, in great quantity. Air.	Bárcena.
"from Guadalajara.	Carbonated gases.[d]	"
Impure"	Organic materials.	"
Quartz.	Liquid or gaseous carbonic acid, pure or mineralized water, *a type of petroleum*, sodium chloride.	Dana.[e]
"	CO_2, N, H_2S, SO_2, NH_3 and F.	Wright.
Agate.	Organic material, oxides.	Dana.
Melanophlogite.	Carbon (1.33): swells upon heating.	"
Hydrophane.	Absorbs molten wax and remains translucent.	"
Granulin.	Absorbs water (17.4 %); is very hygroscopic.	Dana.
Clay.	Absorbs plant nutrients.	Van Bemmelen.

[a] Ibid. 9–16 June. 1890

[b] Chimie organique. CVIII. (t. 1.)

[c] J. Thoulet. Les depots sous-marins. Revue Scientifique. 23 July 1892

[d] Cooked vessels subjected to their action can be assumed to have absorbed something: they acquire a pleasant odor

[e] A System of Mineralogy, p. 188

[g.] *Structures and pseudo-organisms prepared with silicate and colloidal silica.*

Means of preparation.	Results.	No. in Fig. 38.
Anhydrous calcium chloride powdered onto a drop of syrupy solution of sodium or potassium silicate. (Method of Moniez and Vogt.) Then a cover slip is placed on the drop and observed under a microscope (Zeiss 1/8).	Concentric circular structures.	6
	Vesicles with radiations similar to the astrospheres of mitosis.	41–47
	Flagellated pseudo- organisms with or without nucleus. The flagellum is continuous or articulated.	58
	Spermatozoa.	48
	Pseudo-organisms with a flagellum and symmetrical or asymmetrical lashes.	60
	Hyaline tubes, entangled in various ways.	53–53
	Tubes of equal aspect. With little spherical terminals.	22
	Tubes with transverse septa and terminal vesicles, very similar to the nuclear filaments (see figure 33 in this book).	25
	The same tubes, with the appearance of a very large string, in whose interior a liquid circulates, for only a few moments.	36–37
Calcium or aluminum chloride powdered onto a drop of very thick, syrupy alkaline silicate solution. Without a cover slip.	Discs with a nucleus full of corpuscles similar to certain nucleoli in the process of degeneration.	39
	Tubes terminated by irregular appendages.	43
	Radiations, like those shown by astrospheres, formed by very small granules.	35
	Hyaline pseudo-amoebas very similar to natural ones, extending extremely transparent filamentous or dilated pseudopodia, some are only visible if they are tinted with methyl green. Movements of short duration. Interior osmotic flows. Deformations and peripheral oscillations.	1–24–31–32–33–34–40–45–51–54
Almost dry alkali silicate particles deposited on a syrupy solution of calcium chloride.	Amoebae, with dark nuclei. Deformations of short duration.	3–42
Large crystal of calcium or aluminum chloride deposited on a silicate solution.	To the naked eye palpitations and the formation of concentric layers are observed, which	4 7
	slowly advance, like little transparent waves. Tubes visible without a microscope.	
Pulverized aluminum silicate particles and carmine or methyl green. In water.	Each particle of gelatinous silicate seizes the colorful grains and takes the appearance of certain leucocytes loaded with pigments.	12
Aluminum silicate mixed with carmine and coagulated with alcohol at 85 °C.	Very fine granular ramifications. Pseudo-neurons.	8–9
Egg white (which always contains silicates) saturated with quick lime and phosphoric acid. Maceration in water.	Formation of tubes full of granulations at times enclosing a nucleus and refringent nucleoli. After 10 or 12 days the walls of these tubes thicken considerably without losing transparency.	10–11

[g.] *Structures and pseudo-organisms prepared with silicate and colloidal silica.*

Means of preparation.	Results.	No. in Fig. 38.
Crystal of aluminum chloride deposited over alkali silicate.	Formation of an opaline nipple, which grows until reaching a height of 1 or 2 centimeters.	20
Powdered ammonium chloride on alkali silicate.	Lobular structures full of granulations: appearance of asci (reproductive apparati of certain fungi)	21
Sodium silicate treated with aluminum sulfate. Maceration of the precipitate in potash.	After 10 days tracheas, walled and articulated tubes, and cups full of matted filaments are formed.	13–15–16–18
Calcium chloride pulverized over alkali silicate, application of a cover-slip, maceration in water for 20 days.	Important transformations of the pseudo-flagellated organisms, which acquire a radiated nucleus striped nucleoli. Appearance of abundant granulations.	26
Oleic acid deposited over a drop of alkali silicate. Washed the next day with alcohol, ether and water.	Striated muscular pseudo-fibers.	30
Synthetic clay prepared by trituration in a mortar from alkaline silicate and aluminum chloride.	Plasmodia without movement. Insoluble. Imperfect.	
Dried calcium silicate.	Very fine granulated and radiated growths.	27
Weak solution of colloidal silica evaporated at low temperature.	Type of squamae of butterflies, with very fine and iridescent striations, like those of certain species of Brasil (*Morpho*). Tubes and vesicles very similar to the oleates (Fig. 34 in this book, nos. 1, 2, 3 and 163).	28–29
Coagulated colloidal silica.	Squamae of Tisanopterids and hyaline ameboid ramifications.	56–59
Sodium silicate coagulated by phosphoric acid.	Beautiful parallel and denticulate striations.	
A drop of silicate suspended in a cover-slip, applied over another drop of alcohol on the stage.	Very interesting structures. Reticulated and alveolar nipples. Puffs of microbes and spheres that slowly stretch. Extremely small *Streptococcus*. (Zeiss 12/E) in curved regular colonies.	
Silicate and alcohol triturated on the stage during several minutes.	Protoplasmic vacuolated cords. Species of *Gromia* (microscopic organism). Alveolar structure of the protoplasm as is seen in epithelial cells from the tail of the axolotl. These take up methyl green very well.- Rhombic figures full of granulated protoplasm and nuclei are also seen, like those from a dry mixture of phosphoric acid and egg white.	
Syrupy silicate and common ether in excess. Very long contact.	Viscous drops, that later harden; with the appearance of hyaline amebas, which extend bifurcated pseudopods (Amoeba proteus). Vacuolated drops. Multipolar cells, with a refringent nucleus and fine, granulated linear prolongations, like the pseudopodia of Foramnifera.	2

[g.] *Structures and pseudo-organisms prepared with silicate and colloidal silica.*

Means of preparation.	Results.	No. in Fig. 38.
Syrupy silicate and salt water (very concentrated).	Masses of plasmodia and cords with enormous refringent nuclei surrounded by granulated radiations (astrospheres). The next day, other nuclei of distinct size appear. Very remarkable.	8
Potassium iodide powder over alkaline silicate. Without a cover slip.	Nipples with fine granulation. Iridescent sheen. Soluble structures.	
Potassium iodide powder. A few drops of silicate are placed in this.	Hard insoluble ameboid forms, with reticulated surface.	
Lead acetate powder over sodium silicate. Copper sulfate also forms tubes and very fine flagella.	The same as with calcium chloride powder. Nevertheless, globules and nucleated tubes are formed.	
Mixture of equal parts potassium hydroxide and syrupy sodium silicate. Sprinkled with aluminum sulfate.	Slow formation of tubes full of little spheres and wavy filaments, which become articulated in various ways. It is a notable imitation of certain fungi (the white one affecting hops). Very small refringent nuclei, mycelia, conidia, asci.	7

***

The structures vary greatly, according to the dehydration process. If powdered aluminum sulfate is deposited on a drop of very dilute sodium silicate solution,[93] ameboid forms appear, with very fine granulations, nucleus, and palmiform pseudopodia, similar to those that ether produces, but less movable and transparent. (As in nature there are not syrupy solutions of silicate, as was expected dilute solutions give more interesting results.)

Under the influence of common ether, which always contains alcohol, water and other impurities, horseshoe-shaped amoebas often form that move, change their outlines and become filled with vacuoles, and dissolve, like infusoria fragments devoid of nuclei, but if this dissolution is avoided by mixing the ether solution with a little hydrochloric acid, remarkable insoluble imitations appear, composed of a fabric of polygonal cells with very complicated nuclei.[94]

***

In short, imitations of protoplasm are every day becoming more like the natural model, and those prepared with colloidal silica in their structure and their powers of absorption are almost equal to living matter.

---

[93] The immediate application of the cover slip is essential in this case.

[94] We encourage unbelievers to repeat these experiments before denying the facts, which are certified by numerous photomicrographs, sent to various academies.

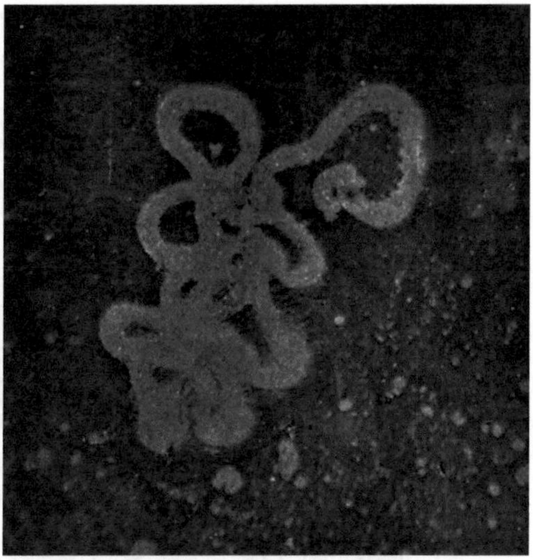

Fig. 36. Artificial nuclear filament formed with silicate and lime that contains egg white. Very slightly magnified.

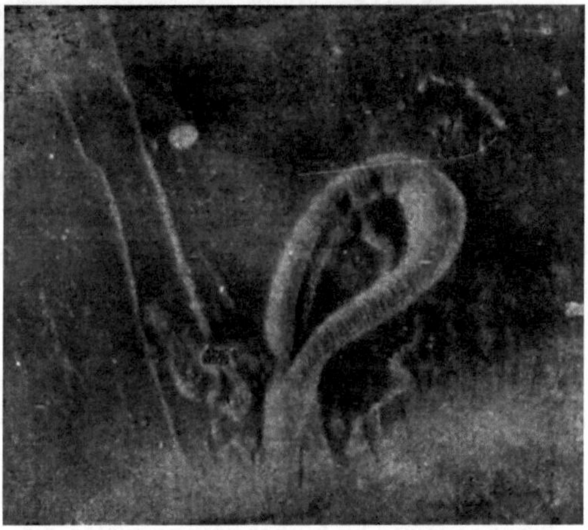

Fig. 37. Colored tubes and filaments formed artificially with the same reagents.

PHYSICOCHEMICAL CONDITIONS REQUIRED FOR THE LIFE OF LOWER ORGANISMS. PROTOCOCOS. TRICODESMIAS. DIATOMS. BACTERIA, INFUSORIA.

Life according to Vogt, began in the sea. "There can be no doubt about this point. Everything found in older layers (Cambrian and inferior Silurian systems), animals and plants, belongs exclusively to the sea and nowhere have traces of a terrestrial organism been found.[95]

[95] Vogt. L'origine des animaux terrestres. Revue Scientifique 15 March 1881, p. 325.

Fig. 38. Structures and artificial pseudo-organisms of silica and silicate colloids. (See text.) These are formed in plain view, instantaneously.

"All animals of the ocean are sustained directly or indirectly by the protozoa and marine Protophytes. According to Dr. W.K. Brooks, the provision of marine animal feed consists of some species of microscopic organisms, which are inexhaustible."[96]

The physicochemical conditions in which precursors of man are found, are equal to those required by higher beings: water, air, heat, reserves.

Just as in the amoebae and other protozoa, the embryo and the human brain, as well as the germinating grain, need a more or less salty, liquid, heat, oxygen and organic and inorganic foods.

For example, *bacteria* or *microbes* are grown in ovens at a given temperature in the presence of free or some combination of easily separable oxygen, gelatin, potatoes or soups, that is, always in moist or liquid media.

The Protococos, so small that 500 would need to be aligned to form a tiny bubble, are complete plants, very simple, without roots, stems, branches, leaves, or flowers: they are reduced to a microscopic sac. And yet, they form a beautiful tapestry of green on the rocks, or a blood red one that extends across polar or high mountain

[96] 1 Jordan and Kellogg. Animal Life, p. 20.

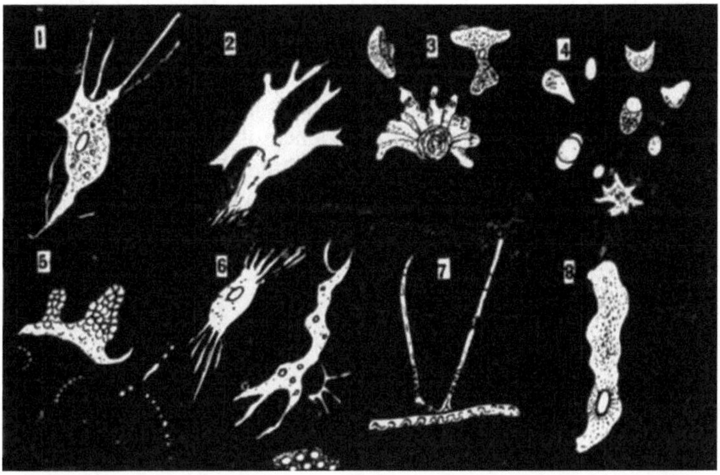

Fig. 39. *Pseudo-organisms and artificial structures*. 1. 2 and 4, sodium silicate treated with common ether, 1, pseudo-protozoan with nucleus and pseudopodia with very fine granules, like those of a marine radiolarian, 2, transparent pseudo-amoebae, imitation of *Amoeba proteus*, 4, imitations of the *Amoeba coli*, to which dysentery is attributed, 3, sodium silicate in dilute solution and aluminum sulfate powder, nucleated ameboid forms, 5. sodium silicate and alcohol at 85 °C; alveolated nipples and microbial forms; imitation of *Streptococcus* 6, sodium silicate and alcohol at 85 °C, nucleated pseudo organisms, hyaline cords, imitating the structure of certain granular protoplasm, 7, sodium silicate added to caustic potash and aluminum sulfate powder: fructigenic mycelia and filaments of a fungal parasite, 8 sodium silicate and salt water: nuclei with granular radiations. All figures greatly magnified (Zeiss. 12/B); these were formed rapidly in view of the observer, in aseptic fluids.

snows, subsisting on, as do more developed beings, whales, oaks, elephants: air, water, heat and absorbed or accumulated food.

Sailors often witness color phenomena produced by microscopic plants that swarm over an area of several square miles. The banks in California have appeared as if washed by seas of blood, and this coloration, especially in the Red Sea, is due to myriads of microscopic algae, the Tricodesmia (bundle of hair), discovered by Ehrenberg in the year 1849. The physico-chemical conditions in which this microscopic vegetation flourishes are the same to which we referred to earlier and certainly have a character of admirable generality, being fundamentally the same for diatoms, plants as small as the Protococci; for all Infusoria, fungi, animals or higher plants, whose protoplasm is and will remain an aquatic organism fond of oxygen and heat.

If, indeed, we examine an organ or a higher being, while overlooking their albuminous or fatty covers of silica or calcium carbonate, if we don't fool ourselves with their sheaths and protection resistances, such as skin, cuticle, bark, chitin, if we seek the first and necessary source of all of this set of structures and armor, we find, inevitably, a viscous substance, the protoplasm, growing, living in a salty liquid (blood, sap, cellular juice) in the presence of oxygen,[97] to an average temperature

[97] Or dissociable oxygenated combinations.

of 8 to 10 °C., absorbing or producing food reserves and manifested by tangible and measurable physico-chemical properties, such as density, surface tension, atomic, molecular and alveolar composition.

Supported by physics and chemistry, the facts and numbers, after a century of painstaking and meticulous research, the science of cytology, has been unable to discover the cherished principles of the life of the cell, and each day gives new proof to the positivistic or mechanistic theory, which alone can lead to a truly philosophical conception of nature.

SUMMARY

Facts of cellular or elementary life.

The protoplasm is an osmotic device, consisting of numerous expandable and permeable alveoli, which absorb water and colloidal bodies through their walls, deforming, sliding, contracting, suffering a more or less considerable expansion.These variations are due to osmotic flow, which can be slowed in oscillating life and sleep, and which are paralyzed by lack of moisture in latent life, but not a definitive way, as in a corpse.

The structural basis of the osmotic apparatus is an emulsion of albuminous substances, according to most authors (or maybe of inorganic nature), and absorbs or secretes a multitude of ternary and quaternary bodies minerals, like fats, albumins and salts, alkalis and phosphoric acid, oxygen being indispensable to all, which determines the exothermic reactions, and water, which transports the materials of the microscopic laboratory.

The general appearance of protoplasm resembles an adhesive, a highly viscous liquid, insoluble in water. In its central portions it contains reserve substances (phosphorylated albumins, nuclein or chromatin), which avidly absorb coloring substances and are necessary for the formation of new protoplasm, distributed in some way between the parts into which the dividing being, after its dimensions pass a certain limit. These phosphorylated substances exist in the male element and the egg, which unite in fertilization, increasing the amount of reserves necessary for the formation of new protoplasm, and new organisms.

This curious substance, the physical basis of life, has been tried to be imitated or artificially formed since 1824, and although to date no one has managed to obtain a complete result, one can be sure that Plasmogeny, the new science of the creation and genesis of protoplasm is solidly grounded on an unassailable experimental basis.

However, as soaps, oleates, tannates, phosphates, ferrocyanates, and silicates produce emulsions very similar to protoplasm, these cannot shed very clear light relative to the chemical composition of the model that has been attempted to imitate, but they can regarding its physical structure, which therefore is no longer considered as an almost supernatural phenomenon. On the other hand, the present status of science the emergence of organic substances, especially albumin, in the bosom of the waters, under conditions of natural generation, cannot be conceived and for this reason a theory has been proposed that provisionally solves the difficulties: protoplasm has an inorganic structural basis, perhaps a silicate, whose elements were provided earlier now supplies the mineral kingdom. The opposite hypothesis seems to be unacceptable and incomprehensible. If albumin were the basis of life and albumin forms itself, it would never have existed. If locomotives have been manufactured by locomotives, the first one could not have existed as the previous one did not exist, and they would never have come to appear before our eyes!

***

Just as at the end of the first book, we shall say now, the summary of facts relating to cellular life: protoplasm owes its activities to known physico-chemical forces. *There is no life force.*

C.

EVOLUTION FACTS

The evolution of the Universe.- The evolution of Earth.-The evolution of minerals. "The evolution of animated beings.-The evolution of man.

GENERAL CONSIDERATIONS

The Ephemera, small insects, live for one day. If we grant them the power to reason about the things found in Nature, the following will result:

Noting that cities are populated by old people, children, adults and young people, whose characters are very different, at least in an apparent manner, the Ephemeral philosopher would conclude that they belong to different invariable species, since in the brief space of one day he cannot winess the development of the child, who will be changed into an adolescence and a youth, and later to an adult and an old person. The dimensions of children and the elderly are so distinct, as are their faculties, countenance, food, language, etc., etc., that the Ephemeral naturalist classifies them into different groups, forming an order for suckling infants, the one of the Toothless, since they have no teeth, and those who possess these appendages, the one of those who communicate by means of an articulated language and those who just scream, the ones that consume only liquid food and the ones that sustain themselves with a mixture of solids and liquids.

The Ephemeral researcher would never guess the close and necessary relationship, that exist between men of different ages, and if some insect, of more prolonged existence, were to propose a theory of gradual evolution and development of our species, he would be mocked by the Ephemera, they would ask of course many tests, but they would not live long enough to examine them.

How would the truth be shown, given that in 24 or 48 hours there is no noticeable change in a child or evolving adolescent? Indirect arguments would be sought, for example, it would be said to the doubters: note, that your species look very alike, all have arms, legs, eyes, nose, mouth (*unity of plan*), which suggests their common origin; note that their differences which are very large at the end points (children and old people) are unnoticeable in the intermediate creatures, there are children who are almost as tall as the young and young people who come almost to the height of an adult. The beard begins to appear in those with 15 or 20 years and gradually increases in the other ages, having all grades of transition between the upper lip of a teenager, the grizzled beard of a man of 50 years and the long one of an ancient octogenarian (*variation, links, intermediate species*); note also that these types of transition are not always aligned like examples in a jewelry store, but there are many missing links, because some children, teenagers and adults, that could serve as a link between extreme series, have died or are elsewhere or cannot be found when needed (*imperfection of the geological record*).

Ephemeral life lasts 24 or 48 hours and that of man for 100 years, i.e. 36,500 times more (*long duration of geological ages*). In short, the Ephemeral-thinker would deduce from their deepest observations with the strictest logic that the child, that the adult and the elderly are and always have been different things, various species that can never fully and radically be transformed.

And if an uncompromising religion increases the darkness of these phenomena attributing them to a creator God who had formed separately the children, young adults and the elderly, the wretched insect naturalist firmly believes to have reached the ultimate truth, defending it with a passion and then presenting the big argument of anti-Darwinists "we do not see that species are transmuted, as the theory requires."

Man, after two or three centuries of truly scientific research, finds himself even worse conditions to form an approximate understanding of nature.

He lives fo 50 or 60 years on average, and in this short period cannot be aware of past events, the gradual transformation of the world of living in 50 or 60 million years, the likely duration of the four previous geologic ages to today.

Ephemeral Life.	Maximum life of man.
One day.	100 years.
Human Life.	Geological ages.
50 or 60 years.	50 to 60 million years.

That is:

1: 36.500 days.

50: 50 '000.000 years.

The first naturalists needed to admit biblical Genesis without dispute and the first evolutionists were neglected and even despised by the most renowned scholars.

Even today Ephemeral Academies exist which deny evolution, precisely because they do not wish to evolve, making wisdom consist of infallibility and absolute constancy.

For us, followers of modern theories, *there was no creation, no Genesis, we are in the wonderful Creation and Genesis, nature evolves, spontaneous generation continues its work and natural selection, powerful divinity, gives the survival to the more capable and differenatiated beings. Seven days were not necessary for creation. Theoretically an eternity would not suffice, for the things of the universe to reach the infinite perfection to which they are predestined.*

THE EVOLUTION OF THE UNIVERSE.[98]

At first glance the universe offers the image of fixity, of immobility. The stars that appear today are those that rose yesterday, and the constellations the priests of Chaldea admired for centuries still shine in space. Poets have sung this aspect of things, contrasting the agitation of men with the serenity of nature.

But with a little care is the mobility of the heavens is soon discovered: among the stars some are seen whose position changes daily: the planets.

Early Greek astronomers, to explain these appearances, devised a series of glass spheres that supported the planets and stars, turning restlessly around the Earth.

Copernicus had the glory of destroying the *geocentric error*, that of our globe to the center of the world. He demonstrated that the Earth is but a planet similar to the others and revolving, like them, around the sun.

Nevertheless, the solar system is not the only one which shows these changes. The stars that are discerned in the depths of the sky are suns like ours and are surrounded by a retinue of satellites in motion.

Nor is this all. It is true that the stars seem to be fixed, but actually they all move about in space. Endlessly, new stars appear and others are hidden. The sun is moving towards the constellation of Hercules.

Among all these systems that the telescope reveals, are embryonic worlds; some more advanced, some already formed; stars that fade and will soon extinguish themselves, and finally, dead and frozen stars. All life stages are represented in the ether, as well as all the phases of agony and death.

"Astronomical measurements have revealed the great fact of the secular displacement of all the stars through the immensity, and we will demonstrate how the constellations are transformed over the centuries. In the old idea of a solid and unalterable firmament, this new idea is a true displacement of heaven."

There are no fixed stars. Each one of these distant suns, shining in the Infinite, is driven by massive movements, which our imagination can barely conceive of.

[98] Dreyfus. L'évolution des mondes et des sociétés, p. 37.

Ursa Major and Minor, will perish, as will Cassiopeia, Andromeda, Pegasus, Gemini, Leo, Orion and all the other constellations and in general all of the population of heaven."[99]

Calculating the movements of the stars of the Great Bear, over extremely large periods of time, almost inaccessible to our imagination, the respective positions of these stars would change to be an almost continuous line in 50,000 years, forming a kind of cross later.

In short, we see that the knowledge of the proper motions of stars transforms our usual ideas concerning the immutability of the heavens. The stars are pulled in all directions through regions of endless immensity, and as for terrestrial nature, celestial nature, the constitution of the universe, changes from age to age, suffering perpetual metamorphosis.

Laplace assumed that initially all solar system materials were dispersed in an immense extension. The matter of the sun and planets was scattered in an immense sphere, which had a radius more than ten times the distance at which the most distant planet, Neptune, is located. To get an idea of this initial state, the mass of the sun and the various planets can be estimated, assuming it is distributed throughout all of this space. It is then found that every cubic kilometer area contains no more than the mass of a peso coin.[100] The lightest mist floating in the air on a spring day cannot give an idea of this rarefaction, perhaps fifty hundred million times larger the remaining air in the vaccuum of ancient pneumatic machines.

And this is not just a hypothesis, a conception of reason, a creation of spirit. In nature there are matters in this state of extreme diffusion.

Beyond the stars, in the depths of heaven, a new type of star has been detected. Here and there white milky masses, in varying forms, such as dim haze, barely condensed and giving a dim light are noticeable through a telescope: these are the nebulae. Some have a surface 640,000,000, 000 times larger than the sun and 12,322,000 million or more than a trillion leagues of gas.

However, our system, the solar nebula also consisted of very diffuse materials, animated by a leisurely movement of rotation. It was a true atmosphere whose limit was the point at which centrifugal force balanced that of its weight. Little by little, as a consequence of gravity, the atmosphere condensed. It abandoned the molecules situated at the primitive limits, then in successive boundaries produced by rotational movement of the sun. These molecules have continued to circulate around this star and others have come into the solar atmosphere, as it condensed.

These areas of successively abandoned vapors formed several concentric rings, which condensed to form planets. Then everything cooled, and the celestial bodies collided with one another, separating new worlds under the action of the heat produced by the collisions. Perhaps these changes have a limit of the radiation of heat into space.

***

[99] Flammarion. La dislocation des cieux. "La Nature." 1875 (2) p. 388.

[100] Very rough calculation.

The above lines seem sufficient to raise the idea that the stars in the universe have not been formed at one time by a single act of creation. On the contrary, as in the world of living beings, creation, extinction and evolution of celestial bodies occurs simultaneously. They are constantly forming in the heart of the nebula, by a kind of spontaneous generation, always possible at the large outer half, the ether, forever fertile!

They are constantly evolving, through periods of a being's life: birth, growth and death. And it could even be assumed that there is a kind of selection, the better constructed stars surviving, those that receive more heat from their suns and nurture a more vigorous population, acting as a gigantic apparatus for the transformation of forces.

***

"The universe has existed from the start in the mechanical state, as a rotating nebula subject to universal attraction, then later heat, light, electricity determined the physical state, the combinations of bodies determined the chemical state, from which naturally derives the ultimate term, the organic state".[101]

In short, according to Crookes, as was said in speaking of the analogies between the protoplasm and the Cosmos "elements have evolved, forming first those with a low atomic weight and later the heavier ones, such as thorium and rubidium."

3. THE EVOLUTION OF THE EARTH.

The Primordial Epoch-[102] (about 500 million years).

The Earth, detached from the Sun is a completely gaseous and brilliant balloon which together with its generator forms a double star – It slowly cools and condenses, it is reduced to a liquid state and becomes surrounded with a dense and stormy atmosphere – It gives rise to the moon. – It continues cooling and condensing, the cold always around 270 degrees below zero, the temperature which reigns in interplanetary space. – It loses its shine, and is slowed enough for the second phase of its evolution, and its surface is converted into a mushy mass.

The Primitive Epoch.– (About 200 million years)

The Earth continues enwrapped in its thick and turbulent atmosphere.– The silica and light metals that float like slag on the mushy surface combine to form the first solid crust of the planet, but internal heating reworks this and profoundly alters it.

---

[101] Dreyfus. L'Evolution des mondes et des sociétés, p. 128.

[102] This synopsis is not mandatory for students.

Primary Epoch.-(Some 6,420,000 years) 42,000 feet.

Reign of trilobites and lycopodiaceae.- Tropical climate.- Large eruptions and rain.	Cambrian period.	Inferior layers are composed of profiles and the superior ones of sandstones and schists.-Azooic in the first phase, in the second rudiments of vegetation and marine animals appear.
	Silurian period.	Represted by sedimentary rocks.-Trilobites prevail.-Cephalopods, fish and some air-breathing animals appear.-Some terrestrial plants appear.
	Devonian period.	The soil is composed of red sandstones, limestones, schists, etc.-Terrestrial vegetation grows.-Trilobites decrease and other crustaceans become more common.-Insects appear.
	Permocarboniferous period.	The lower period is anthraciferous and the the superor carboniferous.-The flora was exuberant, but there were few species.-First reptiles.

Secondary Epoch.-(Some 2,300,000 years) 15,000 feet.

Reign of giant lizards. Mesozoic flora and fauna.- Uniform climate at the beginning, differentiating later.	Triassic period.	Mottled sandstones, shelly chalk, mottled marl.-Molluscs, cartilaginous fish, reptiles.-Conifers, cycads, tree ferns.-Small and strange mammals.
	Jurassic period.	Limestones, marls, schists.-Polyps, polyparies, corals, large reptiles, some chelonians and bony fishes.-The first marsupial. Reign of the reptiles.
	Cretaceous period.	Chalk, marl, clays.-Polyps, foramnifera, toothed birds.-Angiosperms.-Teleosts. Mososaurus.

Tertiary Epoch.-(Some 600.000 years) 3,000 feet.

Reign of the mammals.- Few varieties of flora.- Some important mollusks disappear and others appear, also birds, alsomammals.- Differentiated climate.	Eocene period.	Great depression in the soil.-Clays, sands,limestones, etc.-Nummulites, protozoaria, mammals,quadrumana.- Palms and dicots with deciduous leaves.-Reign of the insectsand gastropods- Giant birds.-Marsupials.
	Miocene period.	Ruminants, pachyderms, Proboscidea, mammoths.- The same flora.-Reign of the birds.-Apogee of the animal world, marsupials will disappear.
	Pliocene period.	Seas and continents take their modern form.-The flora diminishes.-Pachyderms and solipeds appear.-Large terrestrial quadrupeds diminish.-Reign of marine quadrupeds. Modern genera.

Quaternary or Present Epoch.-(Some 100,000 years).

Reign of man.-Flora and fauna become localized.-Great ice ages.-The temperature becomes clement.	Only one period.	Dilluvial and alluvial terrains.-Mammals reach large sizes.-The flora varies little.

NOTE. The duration, thickness and classification of epochs are estimatedin very diverse manners by various geologists and really there is no basis upon which to fix them.

Epoch	**Dominant Organisms**
Primary	Trilobites and fish. Algae.
Secondary	Amphibians, Conifers, Dinosaurs, Reptilian birds, Mammals
Tertiary	Mammals. Deciduous trees.
Quaternary	Man.

The Birth of the Earth.[103]

The birth of Earth began on the day that it separated from the solar nebula.[104]

Animated by a double rotation and revolution movement, there was a vast atmosphere, much thinner than the air we breathe. Little by little, in obedience of the laws of gravitation, the molecules were drawn closer to the center, accentuating the spheroidal shape.

This concentration drove up the temperature greatly. During this first period the incandescent Earth was the site of great events, which are now being verified on the Sun's surface. Huge masses of flaming gas or hydrogen were lifted hundreds or thousands of miles high. Huge tides produced by the action of the sun periodically deformed our globe, projecting boiling masses of lava and gases to the extreme limits of the atmosphere.

The moon was born of these tides one day. Very close to Earth, this in turn produced formidable tides acting on the fluid and viscous spheroid, and little by little, as the rotation of the Earth slowed, our satellite distanced itself.

It is believed two planets our system seem to be still in this formative period, Jupiter and Saturn. The huge red spot of Jupiter, larger than our globe and unchanged from 1878 to 1884, has faded gradually, almost disappearing, perhaps it is a forming continent.

The Earth has gone through these successive phases. At first incandescent, it has begun to cool on the surface, losing heat by radiation. As the temperature of interplanetary space is many degrees below zero, within the bosom of this glacial medium, neither the infall of meteorites, nor the progressive condensation were enough to keep the Earth warm. First spots as the sun appeared, then the outer atmosphere, primarily gas, became a liquid in order then solidified on the surface. The Earth and the Sun, seen from afar, were a double star.

[103] Dreyfus. L'Evolution des mondes et des sociétés, p. 120.

[104] According to Descartes the Earth is an extinguished star.

The various bodies then all existed in a simple state; suspended as vapors, and as the temperature dropped they fell like rain on the ground: at 1,300 °C the rains were zinc, at 350 °C, they were mercury, below 100 °C they were water. Little by little the various elements fit together, forming the present compounds.

Later, in the midst of this immense forge, this gigantic laboratory of Nature, which all manner of chemical reactions took place, a solid crust formed. At first there were some scattered islands in the sea of fire, gradually these grew to meet one another, until they enveloped the entire earth. But the tides that twice a day shook the surface and the gaseous exhalations from the internal furnace, broke the first solid crust no doubt many times.

The primitive globe, because of its rotation, had then taken the ellipsoidal shape it has now, bulging at the Ecuador and flattened at the poles.

This irregularity exists as an undeniable proof of the fluidity of our globe in the first ages. Onto this flattened spheroid different materials accumulated, in order of density, the heaviest near the center, and the lighter ones floating on the surface, like foam.

The inner core or nucleosphere, heavy and metallic, perhaps contained iron impregnated with hydrocarbons analogous to those seen through the spectroscope in the tails of comets.

The first crust was formed from less dense and more refractory bodies, that is, silica and alumina combined with more oxidizable metals; but as it was barely formed, it intercepted the heat emitted by the central incandescent core; the atmosphere cooled and a great number of elements hitherto in a gaseous state, due to the elevated temperature, precipitated on the surface. Thus the first crystallization took place in an environment in which the atmospheric pressure was 250 to 300 times greater than it is today.

The cooling continued, successively depositing volatile elements.

This is the *primitive* environment, which is found at the base of all the geological layers. It shows no traces of animals or plants. It is the *azoic* period in which life had not appeared. The ocean was continuous and without coasts. There were no great depression or elevated points on the Earth's surface.

Cooling took place in the interior as well as in the exterior, the temperature of the inner mass gradually dropped, decreasing its volume.

The outer shell that was molded on the interior nucleus had to follow it in its withdrawal, this was the source of the first inequalities in relief.

These folds or wrinkles gradually grew, putting pressure on the inner mass, the reaction of the incandescent furnace caused tears in the wrapper, the molten mass poured out of them, at first as a paste, later cooling, becoming welded to the walls of the cracks forming the first elements of mountains.

These emerged masses remained separated by depressions where water accumulated, due to the cooling; the major features of geography were drawn, the seas and continents appeared, and the action of water began.

The primitive period has ended. The activity of the material elements undergoes a final separation.

So far, everything had been due to interior energy, the great terrestrial phenomena were nothing like the present ones. The temperature and pressure were many hundreds of times higher than in the present epoch. Life had not yet appeared.

Everything changes in the following ages. The internal energy is trapped beneath Earth's crust, it affects almost nothing in external phenomena, the heat of the surface is due to solar and stellar radiation, if the inner core temperature rises only one hundredth of a degree, the oven expresses its inner existence only intermittently in eruptions or volcanic phenomena which are due, apparently, to partial deposits of igneous materials or lacoliths.

Soon after, life appears and develops through slow and progressive evolution and.[105]

THE EVOLUTION OF MINERALS.

We said that, according to Crookes, all simple bodies are derived from a primordial element: the *Protile* or *electron.*

It is certain that in the first moments of life on Earth, there was only the matter from which known compounds are derived.

The evolution of minerals includes several epochs:

1st. Undifferentiated Protile or electron.

2nd. Simple bodies are formed, Hydrogen, Oxygen, Nitrogen, Iron, Copper.

3rd. Combinations of the first-degree are formed, water (H_20), carbon dioxide (CO_2.).

4th. The rocks and minerals form.[106]

5th. The various agents, water, air, heat, organisms, break down rocks and transform them into other compounds.

Granite comes in enormous masses with no signs of stratification, it does not contain rocks of a different nature and has the appearance of a primitive mass compacted by fusion, then suddenly cooled and crystallized, and can be regarded as the oldest of the rocks.

Later crystallized granitoid rocks were formed, but with an increasingly schistic or slatey appearance. Crystalophilines, gneiss and mica schist, followed, which formed the primitive azoic soil.

Principal primary rocks.

1. Mudstones or shales of the Ardennes.
2. Colored marbles or crystalline limestones.
3. Quartz sandstones and quartzites.
4. Old red sandstones.
5. Anthracite.
6. Bituminous shale. Coal.
7. Modern red sandstone. Metal reefs.

---

[105] The theory of actual causes of Lamarck, Prévost and Lyell, will be discussed later.

[106] See: A. Harker. Evolution des roches igneés. *Science Progress*. No. 1. 1894.

Principal secondary rocks.

1. Sandstones of the Vosges.
2. Variegated sandstones.
3. Shelly limestones.
4. Iridescent Triassic marls.
5. Oolithic limestones and clays.
6. Amorphous limestones and foraminifera.

Major Tertiary rocks.

1. Lignites, clay.
2. Thick limestone.
3. Plaster. Loam. Faluns. Boulders. Sands.

Principal quaternary rocks.

1. Tuff. Sand. Lime. Gravel. Erratic blocks.

***

Biological function has contributed greatly to the constant transformation of inorganic bodies and we can say that the evolution of these consisted in that, although at first homogeneous, they have constantly become more heterogeneous, as part of more and more complicated combinations, and circulating with greater activity, going from immobility and indifference to excessive mobility (beings) and extreme sensitivity.

Evidently water, air and other agents have, in general, more influence on the current complex rocks, than on the primitive ones.

Nitrogen, water and other slightly complex compounds, could be compared to the protozoa and protophytes, while related compounds would be similar to the higher organisms. If these have as a base an inorganic emulsion, the evolution and progressive improvement of the mineral kingdom will seem all the more evident.

Moreover, supposing that these have not changed, remaining in a *status quo*, always seems absurd to us or at least problematic, since everything in the universe is constantly evolving.

One could even speak of a kind of selection because each mineral and every rock variously supports the action of transforming agents such as water and fire; persisting or becoming transformed, under specified conditions.

The studies of experimental geology can mimic a variety of minerals, rocks and formations, by the action of artificial means very similar to those that operate in nature. (Underground sedimentation imitated by Meunier)[107]

According to Lacroix[108] "one of the most interesting questions in the *natural history of minerals, is the evolution of mineral species*. It should not be believed, in fact, that once they arrived at their perfect crystal development, minerals remain unchanged. On the contrary, once formed they are subjected to many physical forces

[107] Géologie générale, p. 230.

[108] Le domaine de la minéralogie. Revue Scientifique. 24 June 1893, p. 772.

that tend to modify, alter or destroy them. None of them can circumvent this law. For example, a sample of British anorthite feldspar, under the influence of water circulating within the rock, loses its aluminum and becomes calcium silicate, wollastonite, the same dissolving action partially epigenizes calcium carbonate (calcite), which in turn is carried off by storm water laden with carbonic acid.

Upon establishing the subtle distinctions between inert bodies and living beings, it was said that a mineral in a vacuum, isolated from natural forces, would remain unchanged for an indefinite period. But even if one could satisfy this condition of isolation, it is not one that dominates in nature and sooner or later a mineral species, no matter how durable, suffers considerable transformation. It can be said that physical and chemical agents never rest not even for granite rocks, that everything moves on the planet and that rest is impossible, as in the interior of a boat floating on choppy waters.

Minerals are apparatuses for the transformation of forces, like living beings, but not as active: this is the only superficial difference that distinguishes them.

THE EVOLUTION OF THE ANIMATED BEINGS

a. Fundamental principle of the theory of evolution

All living creatures have evolved gradually from a unicellular being, through slow variations and selection of the most advantageous in the struggle for existence.

According to this theory the rapid formation of all beings at once is not allowed, on the contrary, it is believed that some have derived from others, being transformed gradually, as happens in a city, which is initially composed of some huts and little by little progresses, serving the same foundations, preserving the primitive neighborhoods, as in Paris, next to the splendid avenues, whose superb, sometimes monumental, constructions, have nevertheless the humble beginnings mentioned: the cabin, the first shelter man could build to protect himself from the elements.

The same happens in the organic kingdoms, in the cities and populations of plants and animals: everything shows a progress, a movement towards perfection and at the same time a humble background which would not be suspected at first sight.

The theory of evolution teaches that the most perfect beings, i.e. those with the greatest number organs adapted to different higher functions, are those who can most easily reproduce and survive in the struggle for life, as in a battle the stronger, more valiant and more agile soldiers are more likely to save themselves, much more so than the cowards, the disabled and the poorly disciplined.

Hence it is that organisms are incessantly perfected and *development, evolution, and progress* occur, making *retrocession* or *involution* impossible for all animated beings.[109]

[109] We deliberately avoid using the term species, because in nature there are only individuals and this term has led to very unfortunate discussions and errors.

b. The three periods of the theory of evolution.

1st Period: Creation. The theory is started with insufficient data:[110]	Moses.—Genesis. 1480. B.C. Linneus.—1700. Erasmus Darwin.—1794. Oken.—1802. *Lamarck.*—1809. Geoffroy.—1830. Goethe.—1832.
2nd Period: Demonstration. The facts of evolution, variation and inheritance are demonstrated.	Lyell.-1830. Spencer .—1858. *Charles Darwin.*—1858 and 1859. Wallace.—1858. Huxley.—1859. Haeckel.—1863.
3rd Period: transformation. The facts of variation and inheritance by physic-chemical forces and the law of nutrition are explained.	*Roux.*—1885 Delage.—1895. Others

1st Period: Creation.[111]

The theory is started with insufficient data:	Moses.-Genesis.—1480. B.C. Linneus.—1700. Erasmus Darwin.—1794. Oken.—1802. *Lamarck.*—1809. Geoffroy.—1830. Goethe.—1832.

MOSES. 1480. B.C.

According to Genesis, the Lord God formed the Earth first, as an inorganic body. Then light and darkness separated, later water and land. This was a planet habitable for organized beings. God then formed first the plants, animals later, first those of the water and air and then those of the land. Finally, he created the human being, the last to arrive, in his image and likeness, to act as the master of the Earth.

According to Haeckel, in this scenario of creation the two most important fundamental propositions of evolutionary theory are shown clearly and simply stunningly:

[110] We cannot cite all of the names of Darwin's predecessors. See the work of Perrier. La philosophie zoologique avant Darwin. Bibliothèque Scientifique Internationale.

[111] To write this historical part of the theory of evolution, we follow the text of Haeckel. Histoire de la Création, p. 35 et seq.

Fig. 40. Lineo. 1700.
Classifier.

the idea of the division of labor or of differentiation and the idea of progressive development, of improvement. It contains two very large errors: the *anthropocentric* – regarding man as the center of creation – and the *geocentric* -considering the Earth as the center of the Universe.

The idea that God formed man from clay, approaches the modern theory of inorganic origin of life.

LINNEUS. 1700 (FIG. 40).

Three thousand years after Linnaeus discovered the way of systematically classifying animals and plants, and his binary method for designating a generic and specific name *initially* facilitated, study and made sure scholars noted the similarities and links between organisms.

Unfortunately, the same Linnaeus proposed a concept of the kind that locked thought in closed circle of iron. Without this unfortunate definition "there are as many different species as distinct forms created by the Infinite Being," the progress of science would have reached an incomparably prodigious height, but on the contrary, today there is still struggle within the Academies, defending and attacking the Linnaean principle of the species!

Linnaeus also accepted the Mosaic legend of the Flood, but, strangely, did not doubt that the union of two different species to give rise to another, new and independent one.

Accordingly, it would be necessary that some species had their origin in divine creation, and others, a mechanical evolution.

In our unauthorized opinion, the existence of the wise Linneus has had a disastrous influence on natural history, as the famous binary system and the supremacy of classification gave rise to thousands of naturalists dedicating their lives to investigations of detail, forming and reforming classifications so that, according to De Candolle, *the classifications need to be classified*, and according to Buffon, the language of science, a horrible language that consists of 300,000 to 600,000 words, is more difficult than that same science.[112]

And indeed, one of the major obstacles that fought the theory of evolution, was the opposition of 3 or 4,000 naturalists dedicated to searching for new species, sometimes qualified with their own last names, and regarded as actually existing and not as transitional forms in the process of continuous evolution: in time, it lost its traditional prestige, diminishing much of the importance of classifiersand their useless collections of curiosities. The situation of humanity and science would have been otherwise if since Linneus' time organisms had been seen as problems to explain rather than as species to be classified.

ERASMUS DARWIN. 1794.

The grandfather of Charles Darwin published a philosophical work in 1794 with the title of Zoonomy, in which he attached great importance to the transformation of animal and plant species for their own life activity, by adaptation to changes occurring in the environment, etc.

It is likely, although not it is not taught in most treatises, that reading this book, which was certainly in the library of the young Charles Darwin, contributed to the germination of the theory of evolution in the mind of this great man.

OKEN. 1802.

The first place among philosophers of nature is usually awarded to Lorenz Oken, who proposed before Goethe, the theory that the cranial vertebrae are the vertebral origin of the skull, and the theory that primitive colloidal substance of a vesicular structure is the basis of organisms, formed in the sea from expanses of inorganic matter during the evolution of the planet. According to Oken, every higher organism, every animal, and every plant, are simply aggregates (synthesis) of these vesicles, infusoria, or cells.

Such ideas, though extremely fertile, were unwelcome and despised because of the absurd form in which they were presented, but are linked in a more intimate fashion with the hypothesis that suggests animals and plants have a single origin, a previous common form.

Oken affirmed that man comes from lower organisms *and has developed, that he was not created.*

LAMARCK. 1809.

The chief of natural philosophy in France, is Jean Lamarck, who in the history of the Darwinian doctrine, is at the first end, next to Goethe and Darwin. To him is due the imperishable glory of having elevated the theory of descent to the rank of an independent scientific theory, and having made the philosophy of nature the bedrock of all biology.

Lamarck was born in 1744, but began to publish his theory at the beginning of the century, in 1801, and developed it in a detailed fashion in 1809 in his *Philosophie Zoologique*. This admirable work is the first reasoned narrative strictly conducted to its ultimate conclusion, the doctrine of evolution. Considering nature organized from a purely mechanical point of view, establishing a philosophical way the need for this point of view, the work of Lamarck by far dominates the mystical

[112] A. L. Herrera. Taxinomistes Heresies. "Les Musées de l'avenir," "La zoologie de l'avenir, etc." Mémoires de la Societe "Alzate."

ideas of his time, and until Darwin's treatise, which appeared half a century later, we find no other book that can be placed next to "Philosophie Zoologique." This book and how far ahead of its time it was will be better understood, if one considers that it was not understood in its time and it remained completely forgotten for fifty years.

Lamarck's greatest enemy, Baron Cuvier, in his report on the progress of natural sciences, where there was room for those engaged to the most insignificant anatomical research, said not a word regarding this essential book. Goethe himself, who was interested in both French philosophical naturalism and "the thoughts of kindred spirits across the Rhine," never cited Lamarck and apparently was not aware of "Philosophie."

The great reputation as a naturalist acquired by Lamarck was not due to this important new work of generalization, but rather to many detailed studies on lower animals and particularly the molluscs, it was also due to a notable natural history of the invertebrates, in seven volumes, published between 1815 to 1822. A detailed genealogy of the doctrine of Lamarck is also found in the introduction of the first volume of this famous work (1815).

The best way to give an idea of the immense importance of the "Philosophie Zoologique," is to quote some of the main proposals it contains:

"The systematic divisions, classes, orders, families, genera and species, and their names are purely artificial works of man. The species are not all contemporaneous, they descend from each other and have only a relative and temporal fixity; varieties breed species. The diversity of living conditions influences, modifies, their organization, their overall shape, the organs of animals, one could say as much from the use or disuse of organs. *Initially*, the most simple animals and plants were formed and then later beings endowed with more complex organization. The geological evolution of the world and its organic population have taken place in a continuous manner, without violent revolution. *Life is but a physical phenomenon.* All vital phenomena are due to mechanical causes, whether physical, or chemical, they have their purpose in the constitution of organic matter. The most rudimentary animals and plants, placed in the lower grades of the organic scale, *were born and are still born via spontaneous generation.* All living or organic bodies in nature are subject to the same laws as bodies deprived of life or inorganic bodies. Ideas and other manifestations of the spirit are mere phenomena of movement which occur in the central nervous system. In reality, the will is never free. The reason is nothing but a high degree of development and comparison of the senses."

According to Lamarck, inferior men derive from apes.

In short, he understood the importance of evolutionary forces and only missed the great principle of natural selection shown by Darwin and his commentators.

Biography of Lamarck[113] Monet, Chevalier de Lamarck, belonged to a noble family originally from Bayern and established in Picardy. He was born in Barentin the 1st of August 1744. After the death of his father at the age of 16, he followed a military career and in Hanover joined the army of Marshal de Broglie, serving there until the end of the Seven Years' War (1763). Disgusted with barracks life, he left the military service and settled in Paris to occupy himself with meteorology

[113] According to Hoefer, the mystical writer whose praises are, therefore, very scarce. "Histoire de la Zoologie". Paris, 1890, p. 303.

and natural history. Housed in a small room "higher than I would have wanted," as he liked to say later, he began his scientific career by a memory "Sur les vapeurs de l'atmospere" and the "Flore francaise," a work that corresponded to one of the requirements of the times and opened in 1779, the doors of the Academy. He wrote the botanical part of the "Encyclopedia methodica" and after Buffon's (his protector) death, he entered the Garden of Plants, as deputy to Daubenton.

Lamarck had taken a distinguished place among botanists when the French Revolution changed the direction of their work. The decree of the Convention (June 10, 1793) which reorganized the Garden of Plants, created two classes of Zoology; Geoffroy Saint-Hilaire and Lamarck were appointed to serve them.

Lamarck had only a smattering of conchology (the study of shells), but despite this he set to work tenaciously and opened his chair a few months later in the spring of 1794 (year II of the Republic). He undertook important work in classification, without giving up his studies, as evidenced by his weather Yearbook (1800 to 1812).

Arago, in his "Story of His Youth" relates a detail worth mentioning: he was appointed to the Institut de France when he was young and was presented to the emperor in the midst of academics who gave the Head of State publications as gifts. After some brief questions which were answered by his neighbors to the right and left, the Emperor approached another member of the Institute.

This was not, says Arago, a newcomer, but rather a renowned naturalist, who had made important and beautiful discoveries: M. Lamarck. The old man presented a book to Napoleon. "What is this?" he asked. "This is your absurd weather alamanac, a work in which you compete with Matthieu Laensberg, this yearbook dishonors your old days, study natural history and I will receive your publications with pleasure. I only take this book out of consideration to your gray hair. Take it!… and he handed it to an aide. The poor Lamarck, hearing these offensive words from the brutal despot, tried in vain to say "I present you a natural history book" and burst into tears.

Many times, the great man in his painful meditations, when he awoke, prisoner to tenacious insomnia, would be happy to repeat Napoleon's harsh words, remembering the unparalleled injustice!

Lamarck was alien to all sorts of intrigue, lived in seclusion, absorbed in his studies, and the education of a large family (he had seven children and was married four times). He was satisfied with a modest fortune, and he refused a professorship created in 1809 at the Sorbonne, because he felt he no longer had the strength to carry it out with dignity.

At the end of his life he became blind and his eldest daughter became his most understanding and unselfish helper. He died on the 18th of December, 1829, at the age of 85.

He founded the important division of vertebrates and invertebrates, publishing a TABLE OF THE ANIMAL KINGDOM *showing the progressive degradation of specialized organs to their extinction*, that is, from man to the polyps and what today are classified among the protozoa.

He said that "it is not the organs, i.e. the nature and shape of the body parts of an animal which has given rise to its particular habits and powers, but on the contrary, its habits, and lifestyle the circumstances in which individuals have found themselves that with time have constituted the shape of its body, the number and the state of its body, in short, the faculties it enjoys." In this it can be seen how important the action of the habits and environment were for Lamarck.

In his "Système des connaissance positives de l'homme (1820) he set out to prove that everything has been formed by nature, with order and this order is serial. In support of this thesis, he reviewed all human knowledge.

In general chemistry he tried to demonstrate that all chemical actions depend on the atoms that enter into the composition of bodies, that these atoms, by their nature, form and disposition, determine the differences between compounds, and in this way he came to the atomic theory (theory of atomicity) and that of definite proportions.

In meteorology he tried to prove that the atmosphere is an aerial ocean, susceptible to more or less violent currents, determined by the pull of the moon in its different phases.

And in fact, atmospheric tides have now been demonstrated.

According to Lamarck, microscopic animals should be considered as natural inhabitants of the atmosphere.

In geology he noted that the assessment of phenomena we can observe, can serve to explain the current state of the world, *whose surface is in a constant state of transformation*. This view was admitted and developed by Constant Prévost and Charles Lyell.

In mineralogy he noted that inorganic bodies are separated from living bodies by a huge space, and they can be arranged in series, either according to the age of their origin or the state of their structure, more and more distant from that shown in living bodies.

In *biology* (the word was created by Lamarck) he established the distinction of recurrent, sensory and peripheral nerves, and of the locomotive nerves of the central nervous system. He recognized that all biological phenomena, from the simplest, absorption, to the highest, thought, are the result of organization.

In phytology he believed that plants are non-irritable, simple or compound, living bodies; they do not form other branch of the living bodies a simple series, but a particular branch, that part of the same point, an inorganic mass capable of organizing, and are, in short, a series in and of themselves. This indicates that Lamarck was not sure of the existence of vast emptiness between the living and the inorganic bodies we referred to above.

In zoology, he first voiced the idea that the orderly distribution of animals should represent the growing series of organization. He was convinced of the doctrine of Epicurus, developed by the poet Lucretius, according to which the production of a body is due solely to the forces of nature.

GEOFFROY. 1830.

Stephan Geoffroy St. Hilaire was the most determined opponent of Cuvier and maintained a memorable controversy with him at the Academy of Sciences of Paris, apparently being bested for want of sufficient evidence to demonstrate the unitary conception of nature, the unity of the mode of organic formation and the intimate genealogical kinship of organized forms.

Fig. 41. Ulfilas Goethe. Poet, philosopher and German naturalist. Cooperated in the theory of evolution, showing the transformation or metamorphosis of floral organs, the spinal origin of the skull, the unity of all organized beings. Year 1832.

In his "Anatomical Philosophy" he expounded the theory of the unity of composition, based on the study of the organs and their relationships and monstrosities, themes we will develop later.

GOETHE. 1832 (Fig. 41).

Ulfilas Goethe, better known as a poet, author of the immortal Faust and the bleak Werther demonstrated the theory of metamorphosis, ascending and descending, of the organs of the plant; considered the skull as a simple transformation of the vertebrae; discovered the intermaxillary bone; he was very interested in the discussions of Cuvier and Geoffroy and defended the theory of evolution, showing that at the bottom of all organisms there is a communal origin and that the differences of the forms comes from the necessary relations with the outside world, revealing a progressive incessant metamorphosis. "Organisms, first confused in a state of kinship in which they barely differed from each other, little by little have been transformed into plants and animals, further developed in two opposite directions to reach, on the one hand, the lasting or stationary tree, and on the other, man who represents the highest degree of mobility and freedom."

2nd Period: demonstration.

The facts of evolution, variation and inheritance are demonstrated:

- Lyell. 1830.
- Spencer. 1858.
- *Charles Darwin.* 1858 and 1859.
- Wallace. 1858.
- Huxley. 1859.
- Haeckel. 1863.

It would be impossible, without being repetitious, to consider separately the immortal and fruitful work of each of these scholars. Suffice it to say, in summary, that Lyell shattered the pernicious theory of geological catastrophes proposed by Cuvier, who tried to explain evolution in this manner, without believing in it, as he supposed that God himself, not content with the fauna and flora of the time annihilated them to form others; Spencer introduced the unfolding theory of organisms almost at the same time Charles Darwin and Alfred Russell Wallace proposed the theory of descent, in 1859, the English naturalist and philosopher T. Huxley and the equally famous thinker, traveler, and zoologist, Ernst Haeckel, of the University of Jena, having supported and guided this kind of preaching.

Life and works of the English naturalist Charles Darwin.

The common man, always light in his judgment, believes the most celebrated naturalist of this century, was limited to saying: "man is descended from apes," doing nothing else worthy of being transmitted to posterity.

It is generally unknown that Darwin traveled widely, studied hard and meditated his whole life on the same problem with unparalleled tenacity and perseverance, publishing a dozen immortal books, full of quotations and observations that reveal at first sight the conscientious work of a man in good faith, a genuine apostle of science, who was unusually well placed by circumstances, who never sold his pen nor had to adapt to any political or religious creed, keeping to death the independence of his fortune and his ideas. He announced what he thought was true, without passion and without interest.

***

Before developing Darwin's theory we should say a few words about the personality of this great naturalist, his life and the path he followed to establish the doctrine of evolution.

Robert Charles Darwin (Fig. 42) was born on February 12th, 1809, in Shrewsbury, Severn, and died on April 19th, 1882, at the age of 73. At 17 (1825) he came to the University of Edinburgh, and two years later to the school of Christ, in Cambridge. He was barely 22 years old, in 1831, when he was appointed to take part in a scientific expedition sent by the British government to reconnoiter in detail the southern tip of the Americas and explore different parts of the South Sea. Like many famous expeditions prepared by England, this was intended to solve both scientific problems and practical issues relating to nautical art.

The ship, commanded by Captain Fitzroy, had a symbolically significant name: the Beagle, that is, the Hound. His journey lasted five years and had the happiest influence on Darwin's intellectual development, and of course, when he first trod the soil of South America, the idea of genealogical theory germinated in his mind, which he was later able to develop in a comprehensive manner. The narrative of the journey, written by Darwin in an interesting style, is far superior to the normal average of these works,[114] discloses the kind personality the author and shows numerous traces of the path he followed to arrive at his great ideas.

[114] A. L. Herrera. La zoologie de l'avenir. Les explorateurs. Memoires de la Societé "Antonio Alzate." 1897.

Fig. 42. Charles Robert Darwin. English naturalist, author of the theory of evolution. He was born on February 12, 1809 and died on April 19, 1881.

The result of this trip was certainly a great scientific report, with Darwin having worked on the zoological and geological sections. He later published a remarkable work on coral reefs which by itself was enough to crown his name with a lasting glory. Most of the South Sea islands are constituted or surrounded by coral reefs. Hitherto no one had been able to explain the singular forms of these reefs and their position on the coralline islands in a satisfactory manner. The honor of resolving this difficult problem was reserved for Darwin; he managed to, noting the activity of the animal builders of coral and at the same time the sinking and lifting the bottom of the sea, which explains the origin of the different forms of reefs.

This theory, like the ultimate theory of the origin of species, perfectly explains the phenomena, attributing it to simpler natural causes, without hypothetically requiring unknown agents.

Among the works of Darwin it is necessary to mention his beautiful monograph on Cirreped Crustaceans, an important class of marine animals, which are similar in their external characteristics to molluscs and which Cuvier had qualified in effect as bivalve molluscs, although in reality they belong to the Crustaceans (convergence or analog similarity).

The extraordinary hardships Darwin endured during his 5 year journey on the Beagle had so damaged his health, that on his return he had to step away from the bustle of London, and since then lived in quiet retirement on his properties in Down, near Brownley, in Kent County, an hour from London.

This departure from the endless turmoil of the great capital was highly advantageous to Darwin, and to this we owe, in all probability, the theory of selection.

Free from the strain of business in London which had made him waste his time and energy, he could concentrate all of his activity on the study of the vast problem. To show what kind of comments had been born in the mind of Darwin in the course of this voyage of circumnavigation, the fundamental thinking of the theory of selection, and how he later completed it, we will quote a passage from a letter addressed to Haeckel, on October 8, 1864:

"Three classes of events made a deep impression on me in South America: first, how species are very similar to one another and are replaced as one goes from North to South, secondly, the intimate kinship of the species inhabiting offshore islands of South America and which are unique to this continent, that I greatly admired, and the variety of species inhabiting the Galapagos Archipelago, immediately to the mainland, thirdly, the close relations that link the toothless mammals and contemporary rodents to the extinct species of the same families. I will never forget the shock I experienced at unearthing the remains of a giant armadillo similar to a present species."

Reflecting on these events, comparing them to others of the same order, it seemed likely that neighboring species may well be the posterity of a common ancestor. But for many years I was unable to understand how they could adapt a similar form to diverse living conditions. I began to systematically study domestic animals and plants, and after a while I saw clearly that the modifying influence of more importance lies in the free choice made by man, and selection of individuals who choose to propagate the species. As I had had studied the genre of life and habits of animals many times, I was ready to form an exact idea of the struggle for existence and my geological work had given me an idea of the enormous duration of time elapsed. Then I read, by happy coincidence, the book of Malthus on "An Essay on the Principle of Population" and thanks to that there arose in my spirit the idea of natural selection."

As seen in this passage, Darwin, since he returned from his travels, dedicated himself mainly and at first, in the silence of his retirement, to the study of domestic organisms, animals and plants; and sure and inevitable means of arriving at the theory of selection.

Both in this study as in others, Darwin proceeded with extraordinary care and attention. Giving evidence of his admirable devotion and circumspection, he did not publish anything about his ideas from 1835 to 1857, that is, for 21 years, nothing, not even a preliminary exposition of his theory, which nevertheless he had formulated in written form in 1844. Without stopping, he accumulated endlessly positive facts, in order not to publish his theory before giving it in an ample experimental base. Fortunately, in the midst of this meticulous work, his quietness was disturbed by one of his countrymen, who in turn, without it being known to Darwin, had discovered and formulated the theory of selection in 1858, and sent a summary of it to Darwin, asking him to pass it on to Lyell, in order that it be published in a British newspaper. This sage was Alfred Russel Wallace, one of the most distinguished and intrepid contemporary traveler naturalists. (Fig. 43).

For many years he had wandered among the islands of Sonda, in the dark virgin forests of the Indian Archipelago, and studying this rich region, so interesting for the variety of its animal and plant population, and had come to exactly the same ideas that Darwin had.

Lyell and the botanist Hooker, who knew his work, decided to publish a short summary, at the same time as the report sent by Wallace. These publications took place in August 1858 in the Journal of the Linnaean Society of London.

In November 1859 Darwin's masterpiece "The Origin of Species" appeared, it developed the theory of selection. However, this book, whose fifth edition was

Fig. 43. Alfred Russell Wallace. Famous English naturalist, collaborator of Darwin. 1868.

published in 1869, was only, according to Darwin, an introduction to the more voluminous and detailed work containing extensive experimental evidence, supported by a mass of facts favorable to his theory. The first part of this great work announced by Darwin came to light in 1868, with the title of "Variations of domestic animals and plants" and contains a rich harvest of data completely demonstrative of the extraordinary changes of organic forms that man can obtain by artificial selection.

The most important consequence of the theory, the kinship of the human family with other mammals was a point that Darwin initially intentionally omitted.

Later, other naturalists established the consequence of genealogical doctrine, and Darwin recognized it in his published work of immense interest entitled "The Descent of Man and Natural Selection."

Finally, the German scholar Haeckel (Fig. 44) gave great importance to Darwinism in his two immortal books: *The History of Creation* and *Anthropogeny*.

LAWS THAT SUPPORT DARWIN'S THEORY[115]

The theory of selection and unlimited variability of the species relies on a number of general facts, acquired by observation and experience:

1st The law of reproduction.

[115] The popular book by E. Ferriere. "Le Darwinism" Paris, 1872, will serve as our guide to explaining Darwin's hypothesis.

Fig. 44. Ernest Haeckel. German naturalist, a contributor to Darwin. Year 1868.

2nd The law of correlations of growth or of organic compensation.
3rd The law of inheritance.
4th The law of geometric multiplication of species and arithmetical multiplication of food.
5th The law of constancy of forms because of the simplicity of structure.
6th The law of the struggle for life.
7th The law of natural selection.

1st Law: reproduction.

All beings transmit life to their descendants in not identical, but varied characters. Thus we see that two brothers differ in height, hair color, the features of physiognomy, gender, character. Fertility is the subject of some general rules. It is inversely proportional to the size of animals and plants. The elephant, rhinoceros, and whale, have a single baby at every delivery, while rats and rabbits produce ten or twenty. The number of fungal spores is very numerous. This rule has many exceptions: for example, a large cod had in its womb 8,000,000 of eggs, being thus much more fertile than the Mosquito (with 200 to 300).

Domesticity strongly affects the fecundity of each species. The pet rabbit can produce up to twelve times per year, while the wild hare, only three or four. The sow has two litters a year, with from 15 to 20 children, while the female boar, the father or ancestor of the pig, has just one litter, with eight or ten descendants.

The length of gestation is in direct proportion to the size of the animal. In the elephant, the largest land animal, it is about twenty months, sixteen for the rhinoceros, twelve for the giraffe, the rabbit takes thirty days. The seeds of ancient trees take months or years to develop. The pollen tube remains in the egg of certain conifers for a year.

The above facts show changes in fertility and the duration of pregnancy, relating to the struggle for life, so that many small species persist only thanks to his numerous offspring, while large ones and those that care for their offspring, such as birds, are less productive.

2nd Law: Correlations of organic growth or compensation.

This was demonstrated by Geoffroy Saint Hilaire and says that the modification of a given organ determines that of another, establishing a compensation, so that the atrophy of one (the eyes of the cave species) coincides with the hypertrophy of the other (the organs of touch).

The organization forms a whole whose parts are related very directly by protoplasmic, nervous, circulatory, and other communications.

The law of the correlations of growth includes all that are possible, shape, functions, appearance, etc. When it comes to particular growth or atrophy of the organs, it is given the name of the law of compensation or organic equilibrium. Goethe said, "nature's budget is limited and a very considerable sum which is used in a certain expenditure is required elsewhere in the economy." When a plant bears many flowers it produces fewer leaves and it is enough to simply cut the wood buds or leaves of a hydrangea so that the economized sap gives a beautiful and abundant flowering.

It is very necessary to look at this law, because it explains very well the development of each organism and the adult characters.

This law has been the foundation of the modern theory of nutrition, as the struggle for life takes place within each tissue and compensation phenomena explain why an organ atrophies or hypertrophies. For example, during embryonic development, man has a tail and other organs that atrophy while another in the anterior region hypertrophies, the brain.

In contrast, idiots, or microcephalics, show many inferior characters similar to monkeys.(See: the law of nutrition).

3rd law: Inheritance (Figs. 45 to 49.)

It is one of the most important in general physiology:

I. Inheritance of acquired modifications.[116]

Acquired modifications are inherited when they exercise a strong influence throughout the body, and are not in the opposite case:

Modifications.	Hereditary	No.
Mutilations, amputations (Circumcision. Hymen).		+
Diseases.	+	
Epilepsy provoked in guinea pigs by hemisection of the medulla.		
Diseases of the heart.	+	
Tuberculosis. Epilepsy.	+	

[116] Many times it is difficult to know if they are inborn or acquired.

Fig. 45. Short-horned cow, example of a sharp hereditary varation, accompanied by many other important changes in general nutrition.- The insignificant variations that influence this, are not hereditary.

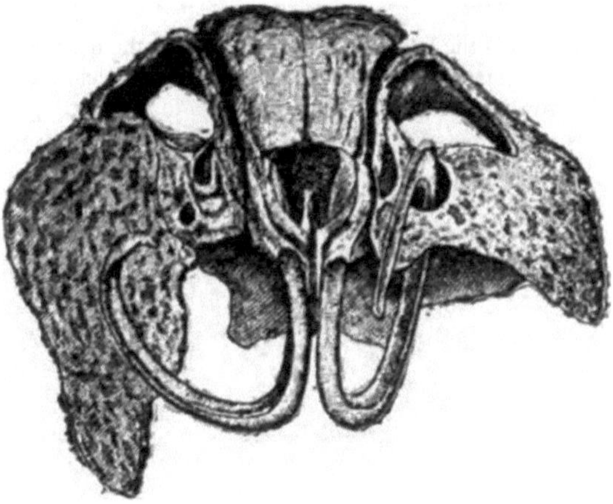

Fig. 46 Gopher or Paca Skull. The upper front teeth have undergone a huge development due to lack of use, caused by the accidental loss of the lower teeth. This hypertrophy is not hereditary. "La Nature." 1887 (2) p. 43.

II. Heredity at the corresponding age.

One of the most important facts of heredity is this: "a peculiarity of organization tends to reappear in the offspring at an age appropriate." As with the size and flavor of the seeds in many varieties of our culinary and agricultural plants, with variations of the silkworm in the caterpillar or chrysalis state, the color of the feathers of the birds or the horns of sheep and adult cattle. Deer Antlers appear when the animal is large, and have almost the same characteristics as the male parent.

***

There are a multitude of other details relating to inheritance. One of the most important is *Atavism* or *retrogradation*, when the offspring do not inherit the characteristics of their immediate ancestors, but rather those of the more remote ones, like a child who resembles his grandparents and not his parents. This is influenced by nutritional conditions, it is sufficient to cool the roots of an oak tree to produce leaves next year similar to those of fossil species of the tertiary strata. It is believed

Fig. 47. On the left, portrait of Mahphoon, the hairy women of Burma, on the right, portrait of her son Moung-Phoset: they had hair of 5 to 6 centimeters over almost their entire bodies. This overgrowth of hair has been transmitted to the fourth generation and is due to an excess of skin nutrition. (*La Nature*. Volume 15, p. 42).

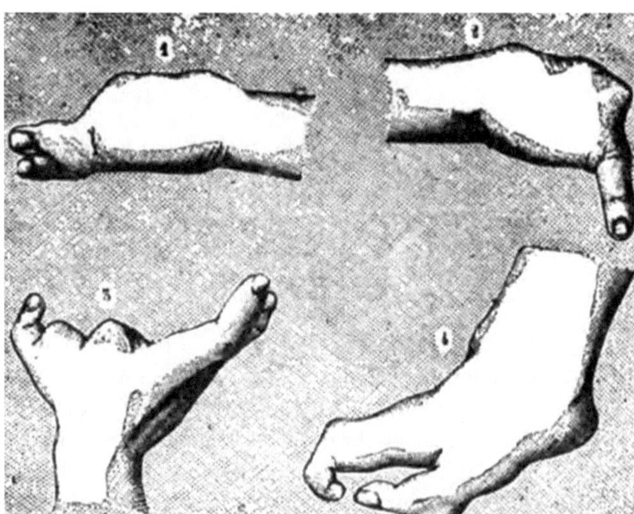

Fig. 48. Monstrosity called ectrodactyly, which is hereditary, and observed in the second generation (According to Verneau). La Nature. 1887 (2) p. 252.

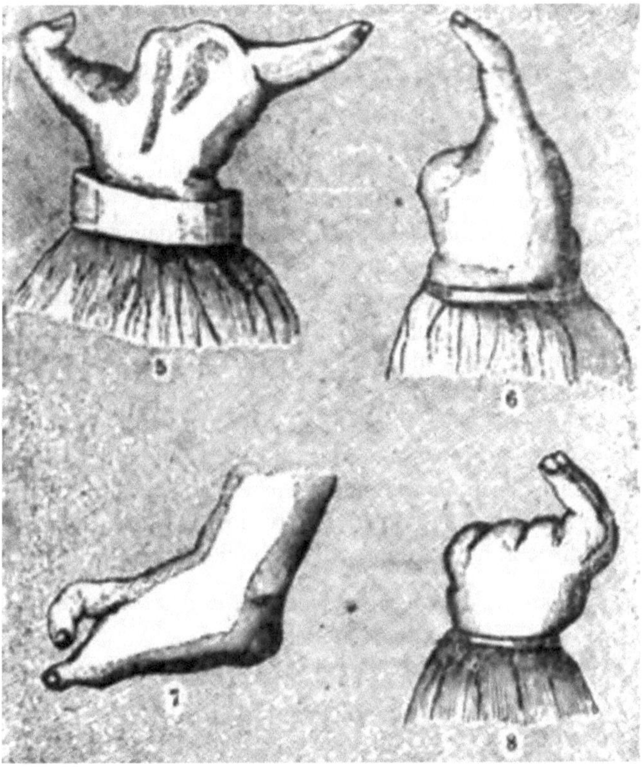

Fig. 49 Ectrodactyly, in the third generation. (According to Verneau). *La Nature*. 1887 (2) p. 252.

that criminal or bestial instincts must be an atavism related to the failure of development of the nervous centers.

The atavism to the primitive forms is frequently seen in domestic animals and cultivated plants, especially in hybrids and mongrels.

***

In any this basic idea should be retained, which has great applications in practical life:

The constitutional defects and large perturbations of nutrition, such as scrofula, phthisis, epilepsy, alcoholism, certain diseases of the heart, cancer, severe nervous disorders, particularly madness, are heredetary or are passed on to offspring, which are stunted and miserable.

A young man, in choosing a spouse, if he has the necessary judgement and rectitude of conscience, should inform himself about the gory history of his fiancée, to avoid the formation of a weak and hapless family. Civil and religious laws tend to this desideratum. Especially hysteria in the mother's history should inspire serious fears that sooner or later the same dread disease will manifest itself in the woman to whom he devotes his affections, and in the end their married life will be difficult.

Teachers, to whom this book is dedicated, have a duty to communicate the preceding idea to their students in the most convenient form, without forgetting the age and circumstances of their students, and preferably to those in their last year.

***

Consanguinity, marriages between relatives, is prohibited by the civil and religious laws and the generality of the savages, which is explained by the almost always fatal result of such marriages, as the hereditary defects increase and almost double in the offspring. But if parents do not have serious constitutional defects or diseases, their children may escape the disadvantages of inbreeding and be robust and have superior qualities. This is what has happened with race horses: the best result from a magnificent pair, and they have not been crossed with individuals from other families (Cornevin).

Parthenogenesis shows that even without the intervention of a male strong descendants may be produced for some time, the intervention of sexual reproduction being needed to correct certain accumulated inherited defects.[117]

The infusoria sometimes reproduce by division, but there comes a time when conjugation is indispensable for improving the condition of the individual.

4th Law: The geometric progression of the species and arithmetic progression of the animals[118]

Calculations and observation show that generally there is a lack of harmony between the fertility of certain species and their food production. This has led to hunger and war and even today it is noted that the major European nations are highly populated and poor in resources.

Elephant gestation lasts almost two years and yet, a pair of elephants under highly favorable conditions, would produce 15,000,000 descendants in 500 years. And what about the fish that lays 8,000,000 eggs!

This law should not be taken to the letter, since any numerical estimate is rough when it comes to the complicated problems of nature. The Engineer Agustin Aragon fought the exaggerated doctrines of Malthus.

5th Law: The constancy of simple forms.

The simpler the structure of a being, the greater the consistency of form and organization and vice versa.

The pace of change is greater in more highly organized beings. Those that have a simple structure show less apparent specialization, and have a smaller number of special organs.

Protozoa are the simplest organisms; the Infusoria, for example, have vibrating tabs that serve many functions, respiration, secretion, movement, etc., while in man there are systems, devices and organs for each function.

---

[117] Here we notice that the microscopic mite destroyer of the Boll Weevil (*Arac-pediculodes ventricosus*. I. Ac.) has been cultivated by the Commission of Agricultural Parasitology for two years and its reproduction has always has been parthenogenetic without significant drawbacks.

[118] Known as the law of Malthus.

Sir Charles Lyell has shown that in two different geological layers, the number of identical species is increased when individuals have a simpler structure. If the same organ can easily adapt to any changes in the medium, its shape will not vary at all. In a mammal each organ has a single function and should vary depending on the environment and the necessities of life, or the animal, as a whole, perishes.

Pancronic species are those that have persisted in many geological layers, "at all times," like some freshwater snails.

(It is difficult to conceive of the persistence of the first organisms after 30 to 80 million years. Perhaps the very simple ones that exist today are later spontaneous generations.)

6th Law: The struggle for life or vital competition.

One of the fundamental laws of Darwinism is the struggle for life. "All animals, Doyer said, are in a state of mutual hostility, at least in this sense: none could subsist without occupying a place a thousand others would seek to take away." You cannot avoid the struggle that each organism fights to preserve its life, whether fending off other beings or the external environment. As soon as a being appears in the world, it begins a necessary, fatal fight.

1. Since then individuals have to fight with the set of external conditions included under the name of *climate*. The other struggle is the one that they hold with each other to find their natural sustenance.

1st Climate. This form of struggle is very difficult. Harsh winters, in some countries, cause the death of 5/6 of birds. Those that survive thanks to their rich plumage and their resistance to starvation, are the only ones who have been able to resist the weather.

In tropical regions, Europeans succumb if they retain their usual living arrangements: they are obliged to take the habits and regimes of the Aborigines. In contrast, in the freezing weather of the North Pole, sailors must ingest large quantities of oil and other fats to supply blood with its vital heat.

When moving northwards we note that certain plants, many of which are rare among many other innumerable ones, grow unconsciously to dominate in cold countries, where they are the only ones, as the others have disappeared. This depends on a slight difference in climate sufficient to make them preponderant, but not too fertile, as to dethrone their rivals and thus owners of the land. Within the limits of permanent snow or deserts deprived of topsoil and moisture, the struggle is determined primarily by climate.

2nd Food. Since individuals grow in almost geometric proportion the food supply is soon insufficient. It is necessary that many organisms perish and only those best at supporting abstinence and hunger survive, or those that in the struggle for food owe their victory to a real superiority in physical force, cunning or ingenuity.

The European bee, recently imported to Australia, is on track to quickly exterminate the Melipona, which is smaller and has no stinger. The mouse has been expelled from a part of its domain by the black rat and this by the gray or culvert rat, in England around 1730 and in France, around 1750, the last of these rats being the strongest and most fierce and especially the most fecund. The black rat is found only in France, on farms and mills away from cities.

In the canals of the Valley of Mexico, a plant imported for fish farming facilities, the romaine or *Eichornia crassipes*, has supplanted many aboriginal aquatic plants, occupying large areas by itself (1898 to 1903).

II. Among the conditions that may influence the outcome of the reciprocal struggle between the species, greater or lower FERTILITY has primary importance, as do the mutual relations of dependency among organized beings.

Benefits of fertility for a species.

Fertility is one of the most effective means to not succumb in the struggle for life. When we reflect on man's amazing consumption of marine fish, among the multitude of those who prey on the inhabitants of the sea, which are more powerful, the persistence of race could not be explained if not for the fact that some females escape the massacre to repopulate the waters.

The same goes for plants. A thousand causes of destruction lurk and decimate, the climate, insects and other animals. They can only survive thanks to their fertility. The common rat does not owe its cosmopolitanism to any other cause, although its omnivorous nature also helps. Germany and the United States, much more fecund than France, are already winning the battle for existence.

This is an example of close relations or symbiosis between two organisms which mutually favors both.

Mutual relationships among organized beings.

Human beings, by nature, are linked by very complex and unforeseen relationships. The simplest accident can result in broad changes.

Cotton is invaded by a weevil, the greater Antonomus, and it has many enemies, but unfortunately the little ants that swarm in the groves take these enemies of Antonomus or weevil, which thrives well, damaging the plant. The main parasite of the weevil is a spider (*Pediculoides*) which also thrives on mason wasp larvae (*Sceliphronus* I. Hy. T.), that sustain themselves on large paralyzed spiders, so that if there is an abundance of insectivorous birds the weevil population can increase by reducing the spider population that eats the larvae of the wasps.

But this compensation is almost always off by one year or another, and this explains the oscillation of all agricultural pests, which follow the fluctuations of their natural parasites and climatic conditions in general.

We would never get to the end if we minutely described the relationships between all things of nature. Suffice it to say that *symbiosis* is the name for the union of two bodies that mutually favor each other (life together) as is the case with the fungus that results in the formation of potato tubercule (Fig. 50) and nitrogen fixing bacteria that lodge in the roots of legumes.

Darwin strongly insisted that most plants have beautiful and fragrant flowers that attract insects (Fig. 52), these being the means of carrying the pollen grains from one flower to another or from one plant to another plant.

Sometimes cross-fertilization is carried out by bats, hummingbirds (Fig. 53), wind, etc., thus avoiding direct fertilization (Fig. 51) which has the drawbacks of inbreeding already mentioned.

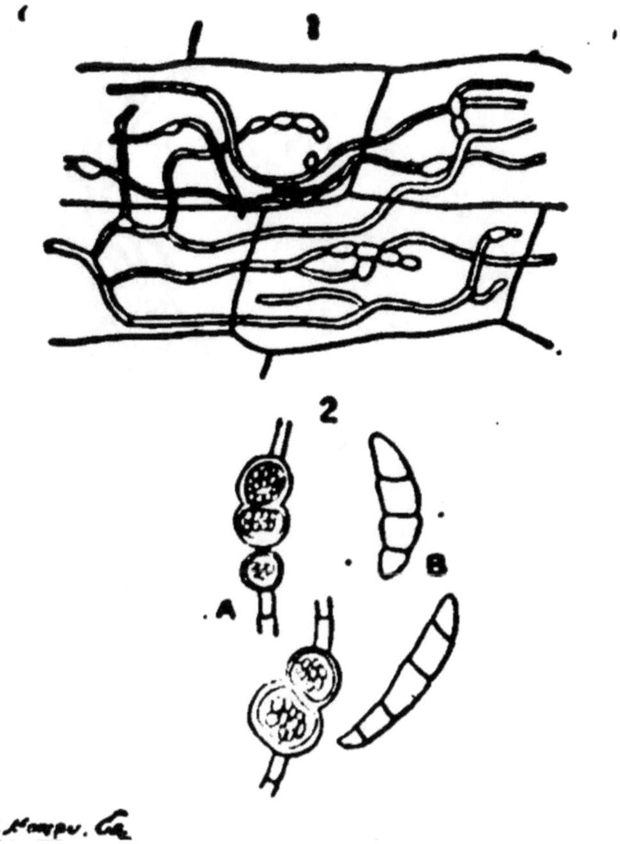

Fig. 50. 1. The fungus (*Fusarium*) which causes the formation of potato tubercules. Amplified 840 times, 2, Fusarium, A, chlamydospores, B, Spores, amplified 590 times (After Coupin. "*La Nature*." 1902 (1) p. 139).

Defense Mechanisms

These are endless and we lack space to describe them.[119] The cell is protected by a membrane and even by poisons and other secretions, such as latex and albumins, which coagulate easily, closing wounds.

---

[119] Cuénot. Moyens de defense dans la serie animal; Herrera. Means of defense in animals. Memorias de la Sociedad "Alzate," 1897. The idea of the dangers and the struggle, the persecutions, the suffering, the spectacles of death, is so ingrained in our brains, that almost all bad dreams have them as their basis. When one dreams of being chased by a bull, whose eyes resemble enormous embers; he who, in his bed is surrounded by ghosts or executioners and bathed in brilliant blood. Barbaric diversions, the Roman circus, cockfighting, bullfighting, are reminiscent of the primitive ages, when war and fight with wild beasts were the dominant thinking and man's principal occupation. Today, intellectual tournaments prevail, but their victims suffer even more cruel pains with what might be called the healing of genius, when the generous impulses have been paralyzed by universal hatred or indifference. (Lamarck, Peysonnel, Watt).

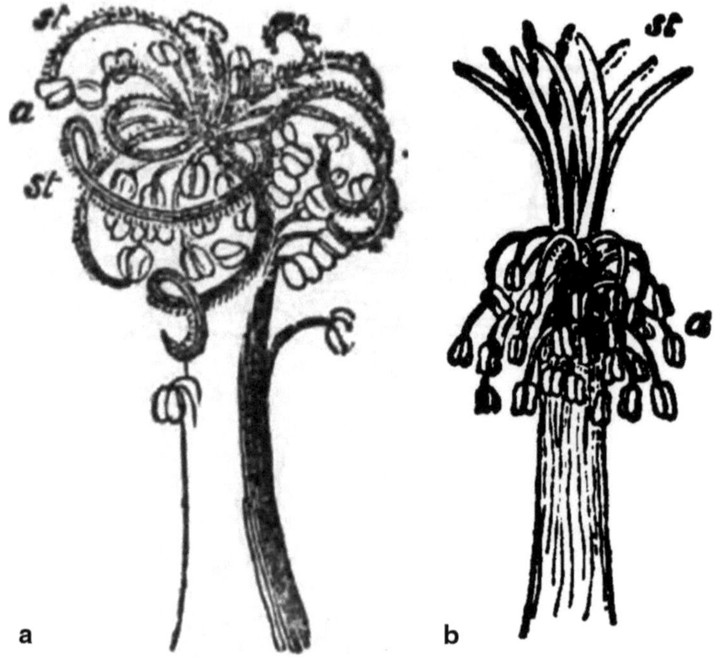

Fig. 51. **a** Flowers adapted to direct fertilization (*Malva rotundifolia.*) The stamens and styles are in contact. After Wallace. **b** Outcrossing by insects: the anthers are below and above the stigmas. After Wallace.

Fig. 52. Cross-fertilization by a hummingbird. The stems and bracts of the terminal part of a cluster of hanging flowers, have become transformed into nectar producers. Birds or insects, upon taking it, receive the sticky pollen on their backs and they take it to another plant of the same species *(Macgravia nepenthoides.)* After Wallace.

Fig. 53. Some marine animals, as shown in this engraving, have luminous or fluorescent organs to attract prey or defend themselves. 1. Fish of the Gulf of Guinea. Its eyes are phosphorescent. 2. Cephalopod collected in African seas. 3. Fish with a fluorescent filament terminal bud, whose light attracts prey. (After to Dr. Laloy. "*La Nature*," 1902, (2) p. 300).

Here is a list of the most common means of defense in the two kingdoms:

Phagocytosis.	Concealment.	Volume increase (frog).
Antitoxins.	Subterranean life (fig. 63).	Unpleasant flavor.
Homochromy or color equal to the medium.	Flight: running, flying, swimming.	In mammals, a number of protective reflexes, nausea, vomiting, sneezing, coughing, smell, taste, and in general respects, and delicate sensibility, etc., etc
Protective resemblance or mimicry (figs. 56, 57.58, 59, 60, 61).	Simulation of death or an active defense (bullies).	Nocturnal, crepuscular, or cave-dwelling habits.
Preserved coloration.	Autotomy or spontaneous amputation (fig. 55).	Colors and other characters, often very visible during flight, which allow the recognition and following of the members of each flock. (Gazelle, fig. 64).
Mechanical means: shells, cuticles, thorns (fig. 62), hair (nettle).	Protective mucus or mucilage.	
Chemical means: odor, poisons, caustics.	Repulsive odor.	
Electric means: discharges.	*Excessive fertility.*	
Phosphorescence (Fig. 54).	*Omnivory.*	
Commensalism and symbiosis: poorly defended species are associated with others better endowed.	*Sociability.*	
	Lack of color or achromy.	
	Shells, nests.	
	Color changes with the seasons, or other circumstances.	

7th Law: Natural selection.

When man wants to create a race that can be distinguished by a specific quality, he chooses (in Latin, *seligere, selectio*, choice) the animals or plants that possess the quality required, and seeks to mate them. By virtue of the law of inheritance, this

Fig. 54. Combat between a crab and an octopus. The crab leaves one of its legs to the enemy and thus escapes death. This iss an example of autotomy. "*La Nature*," 1887 (2) p. 81.

Fig. 55. Stick insect (*Diapheromera femorata*). It looks like a stick, this is an example of protective mimicry.

attribute becomes fixed in the offspring and generally undergoes marked development. This is called *artificial* selection.

Natural selection is based on the fact that the fittest creatures in the struggle for life are more likely to persist.

The first or artificial type is responsible for all of the breeds and varieties of cultivated plants and domesticated animals; the second is responsible for the variety of living species in general, given the tendency of all species to vary, in every way possible, and the heredity that perpetuates these variations, this would have no

Fig. 56. In the lower figure the butterfly has no active means of defense. (*Leptalis orizae*. Piera) and closely resembles the well-defended species of the upper Fig. (*Methona psidii*.Heliconidos) and because of their foul taste is not attacked by insect-eating animals (After Wallace).

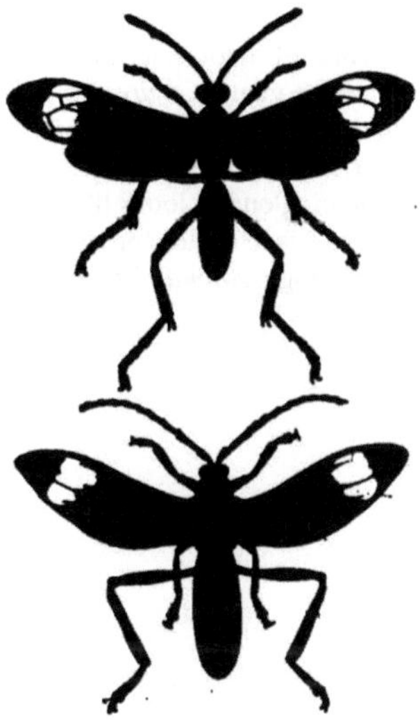

Fig. 57. A mimetic insect, in the lower figure is a beetle (*Coloborhombus fasciatipennis*) having the general aspect of a wasp, represented in the above figure that is well protected, it has a stinger (*Mygnimia aviculus*). After Wallace.

purpose or would give confusing and even harmful results, if selection did not intervene in even the smallest details, keeping only the profitable ones. Without natural selection, disease, hereditary disorders and vices would be perpetuated, and, for example, human societies would become full of dwarfs, giants, madmen, atavists, degenerates, and so on.

Darwin rightly insists that white coloration is very rare in wild animals from temperate climates, because it makes them too visible, exposing them to significant hazards. If in a given species two or more white individuals appear, they will soon be seen and eaten by their enemies, preserving gray individuals or those of a less visible color.

Important observation: natural or artificial selection are not creative forces, and they are not an intelligent and creative entity, as has been alleged by enemies of Darwinism. Selection is a result of the struggle for life and variation.

Fig. 58. Spiders. They have unique shapes and colors. An example of aggressive similarity. Insects approach them without distrust. (After Jordan and Kellogg *Animal Life*, p. 212, fig. 134).

Fig. 59. Mimetic caterpillars of the genus *Cerura*. In the figure above the dormant larva; in the lower, the awakening one: it looks like a horse's face (After Jordan and Kellogg. *Animal Life*, p. 216, fig. 138).

Thus the *cause* and *effect* should not be confused, and in saying that selection has influenced this or that phenomenon, we mean that this is the result of several converging factors, such as climate, nutrition, the fight for females, etc.

Darwin placed great emphasis on sexual selection, suggesting that the most vigorous or more ornate males were preferred by females, but the females really only present themselves and do not choose.

Moreover, these secondary sexual characters are due to males have greater physiological activity than females and females have more need of protection. Thus, the male golden pheasant from China is brilliantly colored and the female is gray, hiding in this way when it hatches. But if the female has an ailment of her ovaries, she grows large golden feathers, increasing the nutrition and accumulation of pigments, which are products of intense oxidation. In summary Darwin's idea can be accepted, as amended, without giving it too much importance.

Fig. 60. Mimetic insects (*Membracids*. Order Hemiptera) that mimic leaf-cutter ants (*Saube*r) of Brazil. The lower figure and the left-hand image correspond to the imitating ant. After Jordan and Kellogg. *Animal Life*, p. 220, fig. 140).

Fig. 61. Sea porcupine (*Diodon hystrix*): the bottom two are swimming in natural position, the top has inflated its stomach and floats at the mercy of the wind and waves, traveling in a passive manner and protected from its enemies (after Jordan and Kellogg).

Result of natural selection.

Divergence of characteristics. Suppose two breeders are trying to breed pigeons, one wants pigeons with very long beaks, and the other ones with very short beaks. According to the familiar principle that no breeder wants intermediate types, but rather extremes, both choose and breed birds with increasingly longer or shorter beaks. What will happen after a few years? The result will be that the successive accumulation of differences will have created two different breeds of pigeons that are highly divergent with respect to that character, which at first glance it does not seem possible have the same origin, especially if, we have forgotten that the

Fig. 62. The gopher and its subterranean lair.

Fig. 63. Soemmering's gazelle. The white spots on the lower and rear parts serve as a characteristic signal or feature, so that are recognizable from a distance and especially when fleeing, by the animals of the same herd (After Wallace).

intermediate types have disappeared, as is the case with breeds of dogs, roses, etc. For Darwin there is no diversity of origin among species but a simple divergence of characteristics.

Extinction of species. Man has almost driven some species into extinction, walruses, whales, lions, wolves, and geology teaches that others have disappeared during different epochs of life on the planet. Thousands of francs have been paid for certain butterflies or birds that are almost extinct.

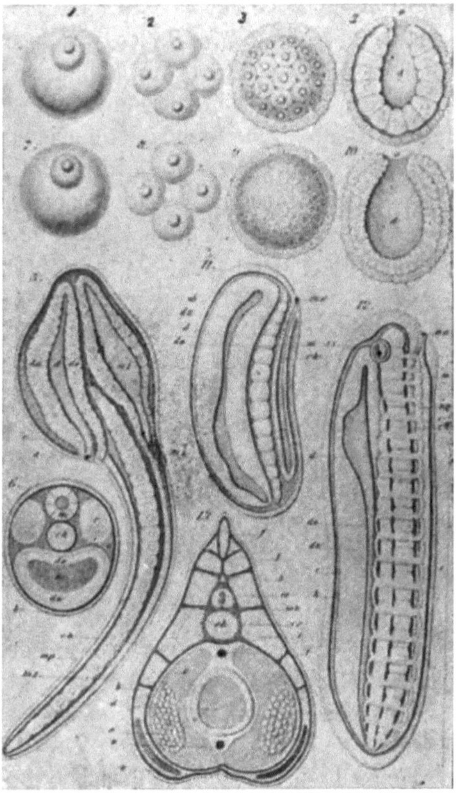

Fig. 64. Embryological ontogeny or development of a *Molluscan tunicate, Ascidia, belonging to invertebrates* (Nos. 1 to 6) *and of a Protovertebrate, the lanceolate amphioxus* (Nos. 7 to 13). Both follow a parallel development from the egg (1 and 7) to the adult (6 and 13) having a rudimentary vertebral column or notochord (ch) (After Haeckel. "Anthropogènie." Pl. VII).

Extinct species do not reappear. The *Mammoth*, the Megatherium, will not be seen by mankind again, because everything evolves, not retrogradation, which is impossible since *the fittest, more differentiated beings, which have acquired a larger number of more perfect organs, survive.*

If a *fossil* species were to appear today it would be overcome by present species.

The tertiary mammals of America had a much smaller brain than their modern representatives.

Finally, it would be impossible for all the characteristics of the child to reappear in the adult, or for those of the protozoan in the quadrumane.

Intermediate geological layers should contain intermediate species.

The Ceralites, mollusks similar to the Ammonites, lived in the Triassic, while the Ammonites inhabited the period immediately above, that is, the Jurassic.

The genealogy of the horse shows the same relationships. (Fig. 82).

In an isolated location the modern species should be descended from the fossil species. Australia is the home of the marsupials, and the fossil mammal found in Australian caves are closely allied with those living today.

The same is true in South America with the Desdentata.

Nature does not advance by leaps and bounds (Natura non facit saltum). If the land could return to life all races and species that it has devoured not only missing a link in the various series would be lacking.

The gap that was noticeable between modern crocodiles and lizards was very great, but Huxley has found intermediate forms or genera: *Belodon* and *Stagnolepis*, from the Triassic, and *Holops*, from the Cretaceous.

In exceptional cases variations are abrupt and very large (for example, short-horned cows), but a reptile has never been seen to suddenly transform into a bird, nor a gorilla into a man.

Unity of Plan.- This is the fundamental similarity found in all organized beings is based on the cell. This similarity is remarkable even in organs that perform diverse functions, such as the bat wing, the fins of whales or fish and the hand of man. For this reason it has been said that *nature is rich in variations and greedy in innovations*. That is, the same organ adapts in many different ways to different needs.

Organic progress. We have said that selection retains the fittest, and to be the fittest more differentiated organs are needed, progress is inevitable, it would not be logical and has not been observed that less gifted beings in terms of defensive apparatus and systems predominate.

Evolution, that is, the progress of nutrition, is inevitable, since it presupposes a multiplication, an increase of forces and parts, and never a decrease of the organs that make the organism more apt to nutrition, by the same fact becoming more perfect. What occurs is what happens in a forest fire, whose intensity grows continuously, or a snowball, that upon falling increases its volume and thus absorbs a greater number of particles.[120]

Humanity is progressing, thanks to the improvement of its institutions, *the division of labor* and human brain development.

Delaunay's Law. Progressive evolution moves from equality to inequality (differentiation) and it is favored by the physiological circumstances that increase nutrition: food, exercise. Regressive evolution is characterized by a decline towards equality and is observed in species on their way towards extinction, the varieties or classes that are degenerate and old.

FACTS EXPLAINED BY NATURAL SELECTION.

I. Geographic distribution of organisms.

The Earth is populated by countless animal and plant species that have been established in various areas, as an effect of natural selection, since not all are or have been equally apt to thrive in the tropics, the circumpolar regions, the sea, islands, lakes, or deserts. This also explains why each continent has its own fauna and flora. However, there are cosmopolitan species such as rats and the darnel, which follow man everywhere.

Naturals barriers such as high mountains, inlets, uninhabitable regions, make passage difficult for organisms which move slowly and do not fly, such as frogs. Physiological barriers are even greater obstacles. For example, or (A'pitos) or water

[120] A. L. Herrera. L'origine des individus. Memorias de la Sociedad "Alzate." T. XI, 1897–98, p. 140.

hares, a species of duck which can no longer fly and which walk very awkwardly, are permanently limited to lakes that connect or were connected in another era with a river.

Frogs do not exist in oceanic islands because the salt water kills them and the only mammals found there are bats. But if the island is close to a continent the inhabitants will be more numerous and varied.

Wallace and other naturalists have proposed geographical areas, the Nearctic, the Neotropics, etc., which have little interest, which only exist in the mind of scholars, as the law of reproduction and the struggle for existence require the dissemination in all directions, of almost all species through time and often in spite of obstacles.

Active emigrations such as those of a Swallow fleeing winter, or *passive* ones, such as those a snail traveling on a floating log may also be *periodic* (Ganga) or *irregular* (Filomena) and contribute effectively to the mixing of fauna and flora. For example, coconuts and sea porcupines cross the oceans, floating on the waves. Man, through his continuous migrations has come to invade almost the entire Earth.

Many species that inhabit different areas today come from a common center. Among the animals living together in Attica, some have migrated to the North: the Beaver, the Reno, the Lemming, and the musk ox; others are only found in tropical Africa: the hippopotamus, the spotted hyena, and the African elephant. This coexistence in Greece, of animals living in such diverse countries (Lapland and Africa) is a fact of high importance. It is believed that there was a glacial period, and the cold expelled many species, which quickly changed their homeland.

11. Rudimentary organs.

Nothing is more common in nature than the presence of rudimentary organs.Mammary glands or breasts in males.

A lung in snakes.

Pelvis and legs in the same.

Teeth in whale fetuses,

Fused wings in island insects.

Wings in ratites (Aves).

The semilunar replicate in the human eye.

The cecal appendix in man.

It will be understood that these organs have atrophied at the expense of more necessary ones during the struggle for life. In contrast, microcephalics, idiots, criminals, and prostitutes have many atavistic characteristics which do not correspond to their sex, because their entire organism is imbalanced.

APPENDIX TO THE THEORY OF DARWIN

Critique of current classifications.

Darwin shows that all classifications are arbitrary and artificial and only genealogical classification is acceptable, although it is impracticable, and as the fossil series is incomplete, the result is the aforementioned arbitrary practice.

Naturalists should never forget that *things and organisms have many fundamental analogies that make an absolute separation of classification impossible, but their superficial differences necessitate a provisional classification to facilitate study.*

It really is impossible to classify the way scholars would like to, since all beings are composed of cells and these vary only in minor details. Chemical classification would be even more difficult, as all protoplasm shows very close analogies and all of it contains fat, proteins and salts, in various proportions.

It has been possible to graft certain organs of one animal onto other, forming artificial monsters by fusion. How can one classify under these conditions what is and will always be identical, essentially identical!

I. Analogous and adapted characteristics.

It was long believed that the particularities of organization, which determine the life habits and the environment in which each organism lives, should have great importance in classification. However, nothing is more false. The common man regarded whales as fish, because they resemble the superficially and they live in water. But whales are cousins (removed 10 or 12 times) of man, belonging as does he, to the class of mammals, while fish belong to the last. Whales breathe through lungs, like men, are warm blooded like men; the whale gives birth to one calf and breastfeeds it with her two breasts, like a woman with her baby. The whale is animated by the most tender and passionate love for its offspring, if a whaler comes to seize her calf, the mother suffers such delirious love, running to get in front of the harpoon, in a kind of suicide. Because of this and other examples cited above, analog similarity or convergence, destroys the argument founded on characteristics of analogy and adaptation, it would be nonsense if all the animals that fly, swim or converge in another sense (the bat and the flying fish) were classified together.

II. Organs of supreme physiological importance.

The reproductive apparatus and the nervous system vary only slightly, but cannot form the basis for a complete classification *of the species without organs*, and in any case only allow the establishment of large groups, without accommodating certain links, such as *Amphioxus*, that are a step or transition between vertebrates and invertebrates.

The classification of the Coleoptera is based on the number of articulations of the tarsi and there is a naturalist who has spent his life studying the characteristics of the legs and antennae of the Weevils, publishing the result of his observations in a work of seven volumes!

III. The embryo.

Milne Edwards and Agassiz believed that the characteristics of the embryo can be used to build accurate classifications, but they did not understand that this gave a proof of the theory of descent, since embryos pass through stages similar to those which the family or class they belong to has gone through (the law of biogenetic or embryonic similarities).

Conclusion.

Natural system is a contradictory expression. Only genealogical classification may have any accuracy. The terms genera, families, orders, and species express nothing but various degrees of difference between the descendants of a common ancestor and are arbitrary conventions, estimated in many different ways in different countries, by different authors and in different eras.

There are no two equal classifications made by two different naturalists.

It was stated that the species is very similar set of individuals that can reproduce and give fertile descendants, but hybrid families of mules have been obtained and that definition is inapplicable in practice and although it was founded in details of fertility or infertility, these have little importance. We have seen that pointy sperm can fertilize the eggs of various species of amphibians. We do not believe that a whole school of naturalists can rest on such flimsy grounds and that that are less stable every day.

In nature there are only individuals and even the notion of individuality is difficult to relate with the caracteristics of lower animals, which form colonies similar to individuals or individuals similar to colonies (Fig. 73).

VARIOUS TESTS OF THE THEORY OF EVOLUTION

Ontogeny or individual development.

Haeckel has written great and profound dissertations demonstrating that ontogeny (individual development) is a kind of recapitulation of phylogeny (the development of family,

order or class.) Embryos of different animals resemble each other more the less their development, all begin as a simple nucleated cell (fig. 65, no. 1) which rapidly divides (Nos. 2 and 3), which later forms the gastrula or primitive stomach (Nos. 4 and 10 in Fig. 65 and Fig. 69) that in some zoophytes is an organism,

free-living larva (sponges, Fig. 69). The human embryo (Fig. 68, M) goes through different phases that resemble a protozoan, an amphibian, a fish, having gills, a tail, webbed digits, which later disappear. According to Haeckel, "every animal, every plant, reproduces in quick succession and in general contour, the long and slow evolutionary series of transitional forms through which their ancestors passed since the most remote ages.[121]

Metamorphosis. Species subject to metamorphosis manifest similar phenomena. For example, the halibut or flounder in its early days is symmetrical as all fish are, and little by little, one eye migrates until it is next to the other (Fig. 71.) Crustaceans pass through curious larval stages (zoe, megalops, Fig. 70) which correspond to primitive forms (See also the metamorphosis of a fish, Fig. 72).

[121] Histoire de Création 18, p. 359.

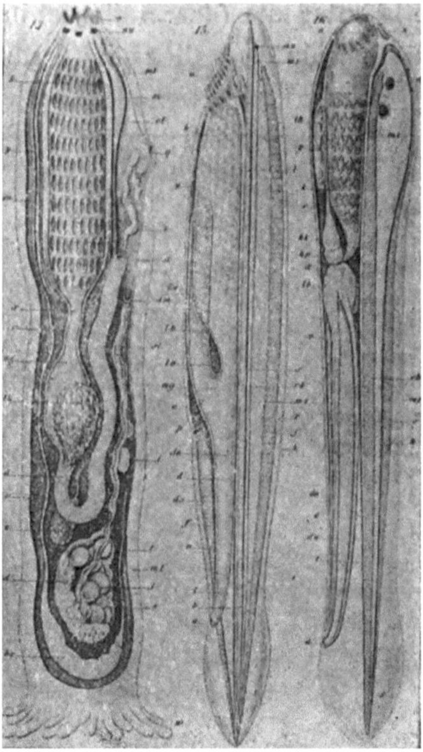

Fig. 65. Extraordinary anatomical analogies of *Ascidia* (14), an adult Amphioxus (16) and a lamprey larva (16) (After Haeckel. "Anthopogènie." Pl. VIII).

The origin of colonies. The study of protozoa and in general of the zoophytes has come to demonstrate, according to the work of Perrier, that higher organisms are colonies of protozoa that have gradually become associated with the goal of mutual protection (see Fig. 73). We have said that the notion of the individual rests on an insecure basis: polyps sometimes consist of well-differentiated individuals, some nutritive, others reproductive, still others guards: but all communicate with a general cavity in which fluids circulating food.

Polymorphism. As species have no absolute invariability or fixity, they show variations and a multiplicity of form or polymorphism, depending on the circumstances of nutrition. The leaves of plants sometimes vary widely according to the impact of external forces, especially light (Fig. 74).

Variation. Darwin wrote a book in two volumes showing that all beings and all of the organs of each vary incessantly (Figs. 75 and 76) and that the stale ideas

about the fixity of the species are baseless. Recently, in the United States, very tired and detailed studies regarding changes in size, color (polychromism), etc. etc. have been made.

Adaptation.

As species are not immutable, they can accommodate themselves to different living conditions, which has happened to man and domestic animals and plants. This adaptation requires that the protoplasm is modified, as occurs in unhealthy regions, and

Fig. 66. This Fig. and the following one show the analogy of form that exists in the early stages of development, including in the human embryo and in that of other vertebrates. This analogy is all the more perfect the less evolution has advanced, and continues the more the adult animals resemble eachother F., Fish. A, Salamander. T, Turtle. H, Chicken (After Haeckel. "Anthropogènie." Pl. IV).

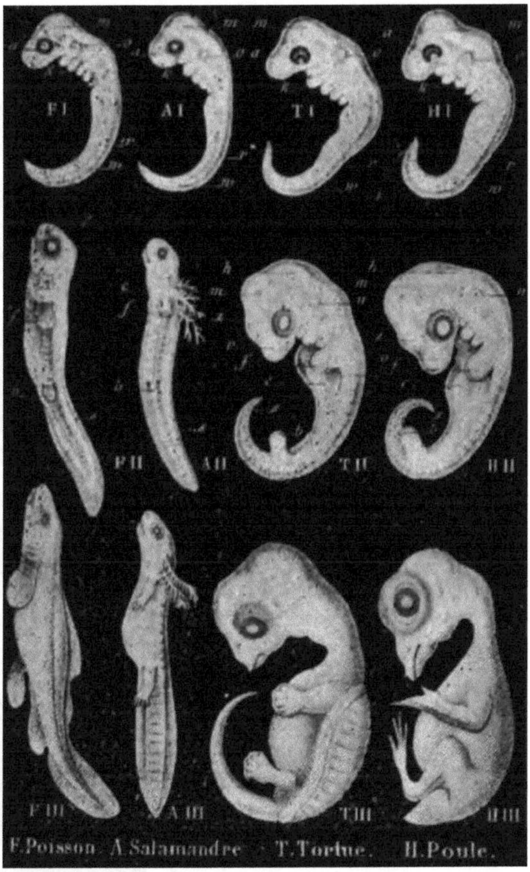

even the organs become highly modified (Fig. 77) adapting to new needs. Everyone knows that there are people without arms that sew and write with their toes! If the modifications affect general nutrition, they will become fixed by heredity.

Analogous similarity or convergence.

Similar needs produce similar adaptations. Australian marsupial mammals have undergone changes similar to those of the whole group around the world, and there are Marsupial herbivores, rodents, primates, carnivores, etc.[122]

Division of labor, the basis of improvement.

The struggle for life leads to specialization and the division of labor. Just as the abundance of doctors has led some to pursue a specialty, the same occurs in nature as the organs of each species and the species themselves become specialized, improve, divide work, as occurs in social insects (Fig. 79), whose members have determined functions, some are devoted to the defense of the colony (termite soldiers),

[122] A. L. Herrera. Catálogo de la colección de Mamíferos del Museo Natural. Second Edition (several examples of convergence).

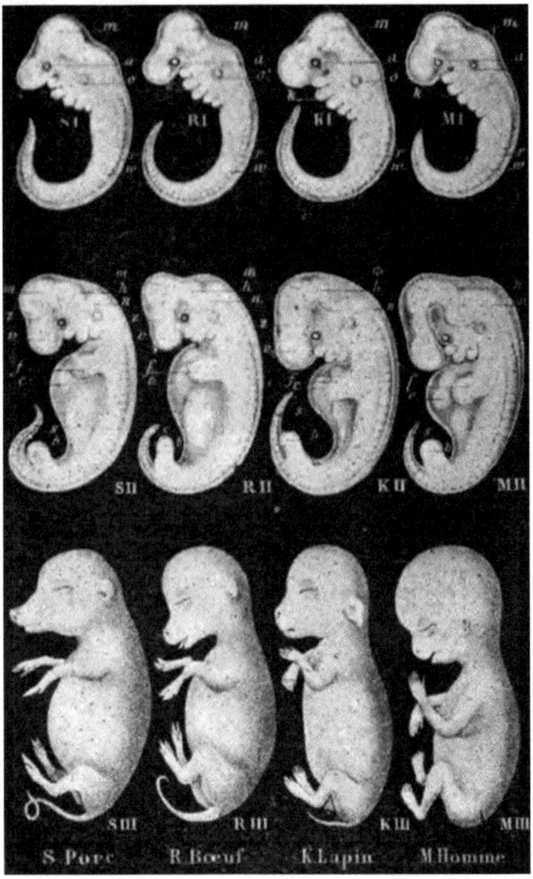

Fig. 67. This shows the state of the embryos of a Pig (S), a Bull (R), a Rabbit (E) and a man (M) in the 1st, 2nd and 3rd week of gestation; v, forebrain; z, midbrain, m, midbrain; h, hindbrain; n, postbrain, I., bone cord, and, nose to, eyes, or, ort? ja, k, gill arches; w, spine, f, forelegs, b, hind limbs, a, tail (After Haeckel. "Anthropogènie." Pl. V.).

others the work of food and construction (wingless ants, bees) and others, finally, to the conservation of the species (queens).There are cases where the male is stunted and produces only invigorating germs and the female accumulates food reserves (Figs. 79 and 80).

Really, the separation of the sexes, which starts in lower organisms, is a great improvement, thanks to it males and females can separately address their needs and work. Females are the nutritive treasure of the species.

Parasites. Degeneration.

Some organisms remove themselves almost entirely from the struggle for life, living as parasites at the expense of more active beings, then degenerate (Fig. 81) and lead a passive and miserable existence, losing the organs of movement and the sensation (only). These can be regarded as fossil species in danger of extinction.

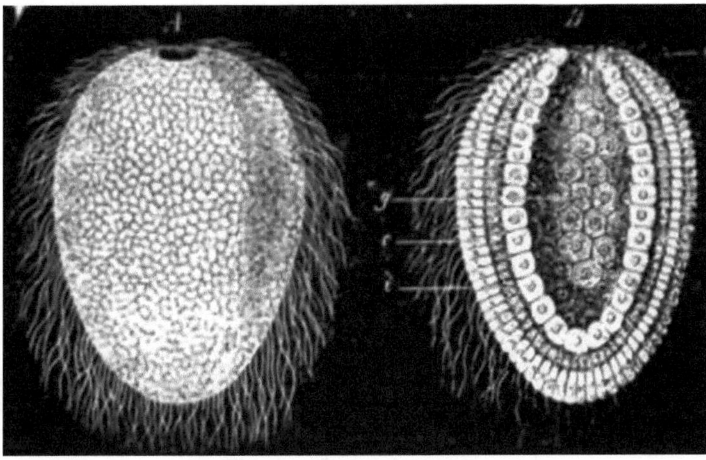

Fig. 68. State of the gastrula or free embryo of a calcareous sponge, after Haeckel. Many embryos of different organisms pass through a similar state.

Fig. 69. Metamorphosis of the crab. a, zoe state, b, megalops state, c, adult state. (After Jordan and Kellogg).

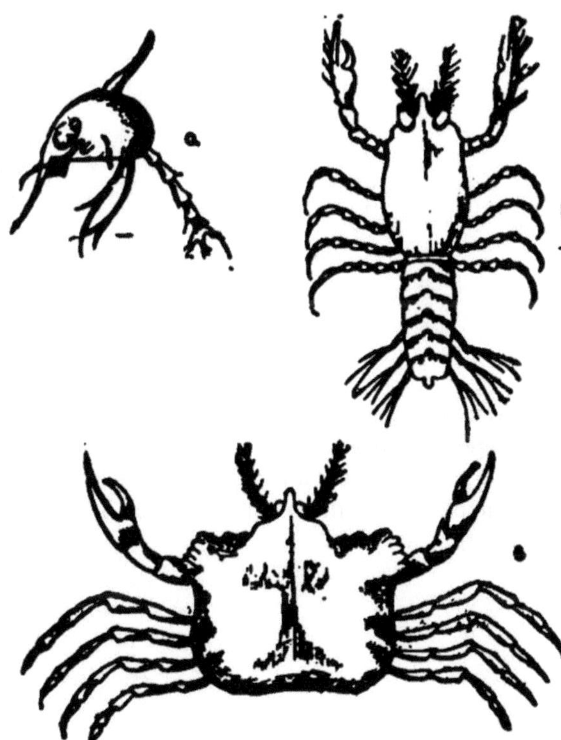

Fig. 70. Development of the Sole or Flounder. The young animal's eyes are symmetrical and bilateral, then migrate to one side.

Fig. 71. Postembryonic development of the dace (*Albula vulpes*) showing metamorphosis (After Jordan and Kellogg). The larva is a free embryo.

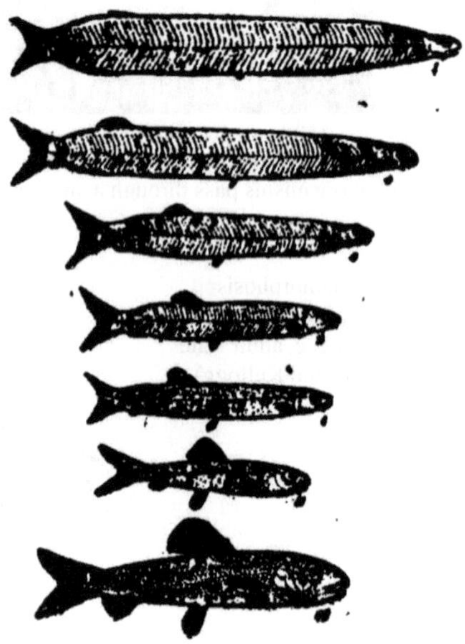

However, all organized beings are perfected constantly, left behind if the majority of stragglers and degenerates, which are left behind in the struggle for survival.[123]

Some parasites cause a very important selection in resistant varieties and species (for example chahuixtle on wheat), especially when they destroy the sexual organs. (A. Giard).

[123] We cannot discuss here the issue of degeneracy and parasites. See: Darwin. L'origine des espèces, p. 290.

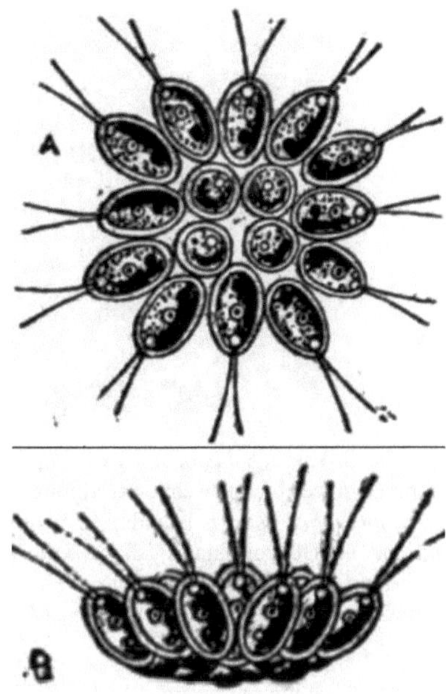

Fig. 72. Higher organisms are colonies of protozoa. This is how this comes to be observed in some species, as represented in this drawing (*Gonium pectorale*). A. colony seen from above, B, profile view (After Jordan and Kellogg).

Fig. 73. Polymorphism in the Euphrates poplar. Four leaf shapes. (After Wermael. "La Nature." 1887, (2) 215).

THEORIES OF MENDEL AND DE VRIES[124]

Elementary species and varieties themselves owe their origin to essentially different changes in the hereditary traits of organisms. These changes occur in an abrupt manner and de Vries calls them *mutations*. These elemental species are born through

[124] Not mandatory for students.

Fig. 74. Locust (*Schistocerca*) of the Galapagos Islands, Pacific Ocean, all descended from a common ancestor, but scattered among various islands and varying in size and color. *Schistocerca melanora* (Carlos Islands); b, *S. intermedia borealis* (Abingdon and Bindloe Islands); c, *S. intermedia* (Duncan Island); d, *S. literosa* (Chatham Islands); e, *S. melanora lineata* (Albemarle Islands); f, *S. melanora immaculata* (Indefatigable Island.) The species *intermedia* is probably a hybrid between the other two species. (After Jordan and Kellogg.)

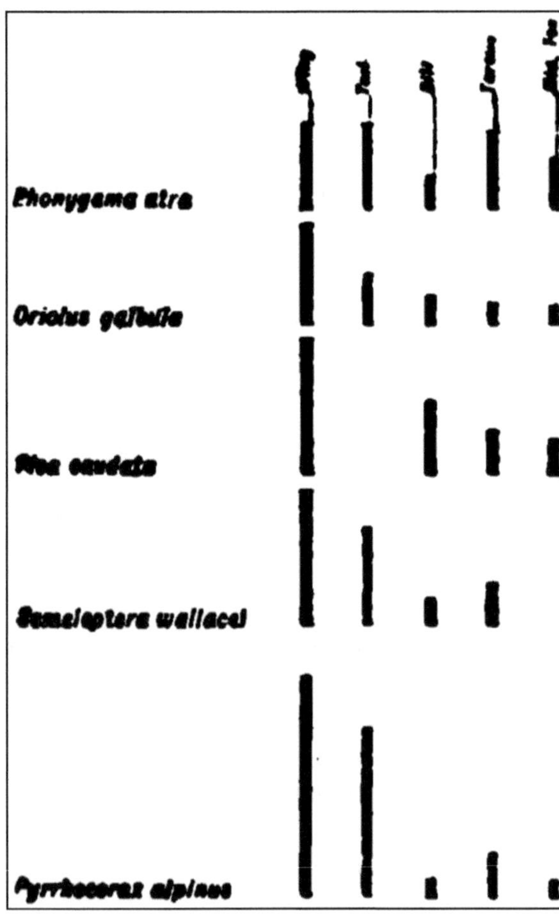

Fig. 75. Graphic representation of variations in size. In the first series of vertical lines, on the left, of wings, in the second, of the tail, in the third, of the beak, in the fourth, of the tarsi, in the fifth, of the middle toe. The birds measured are: top, in the first line, *Phonygama atra*; in the second, *Oriolus galbula*; in the third, *Pica caudate*; in the third, *Semeioptera wallacei*; in the fourth, *Pyrrhocorax alpinus* (After Wallace). This example is one of the many that prove the variability of organisms. There are no fixed or final forms.

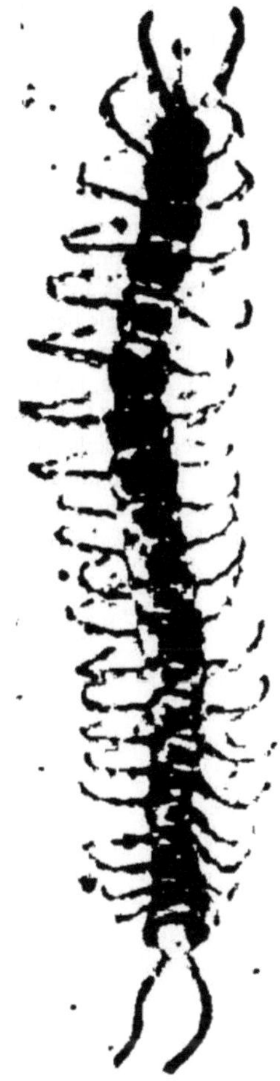

Fig. 76. Centipedes. The first pair of legs is modified and serves as a clamp for grasping and biting. This is an example of adaptation of locomotion to defense and to the capture of food.

progressive mutations which add a single characteristic or a specific unit or units to characteristics already present. Proper varieties, in contrast, are not formed by the addition of new principles, but by a change in the state of activity of existing principles.[125]

(In our unofficial opinion this theory has no basis, since no one has been able to establish the definition of the species and of variation. In nature there are only individuals. How do we apply the theory of mutations to that which exists only in our imagination? How do we apply that theory to the numerous forms of plants and domesticated animals, and to the 192 British plants which according to H.C. Watson are considered varieties, but certain naturalists have elevated to the rank of species? Babington says there are 251 species and Bentham 112!)[126]

[125] Botanisches Centralblatt. No. 13, 1903, p. 275.

[126] Darwin. L'origine des espèces, p. 51

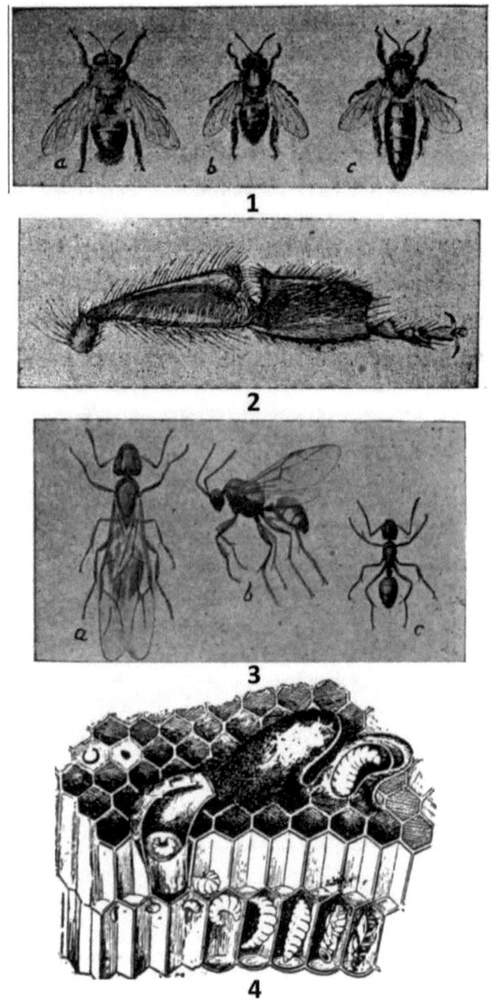

Fig. 77. Adaptation and polymorphism in social insects. 1 male or drone; b, worker bee or sterile female; c, queen or fertile female; 2 Hind leg of a bee. The concave surface of the upper articulation and its hair, are for recovering pollen, the pollen collectors are the sharp edges of the angle between the two large segments of the leg 3. Ants: a, female; b, male; c, worker (*Camponatus* sp); 4, beehive cells with eggs, larvae and pupae: the large ones are the queens. (After Jordan and Kellogg.).

The law of Mendel or of dominance. When two animals or plants that differ in some characteristic are mated, all of the descendants frequently show the characteristic of one parent, which is called the dominant characteristic. For example, when gray mice are crossed with white mice, all of the children are gray. The other characteristic is called *succesive* because it is latent.

This law does not apply to hybrids, which have mixed characteristics, but which produce germ cells that carry *only one* of each pair of characteristics by which their parents differ, such that in the second and subsequent hybrid generations, a number defined forms, in defined numerical proportions occur[127] (purity of germ cells).

[127] According to W. E. Castle. Science. Vol. XVIII, p. 406. Sept. 1908, and G. Cuboni. Le leggi dell'ibridismo. Rome, 1908.

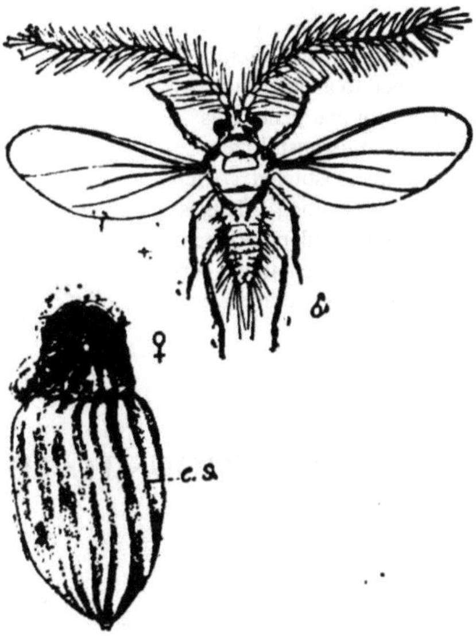

Fig. 78. Sexual dimorphism in an insect (*Icerya*). The male has wings and huge feathery antenae, the female is sedentary, does not fly and has is covered in wax.

Fig. 79. Sexual dimorphism in *Syngamus*.

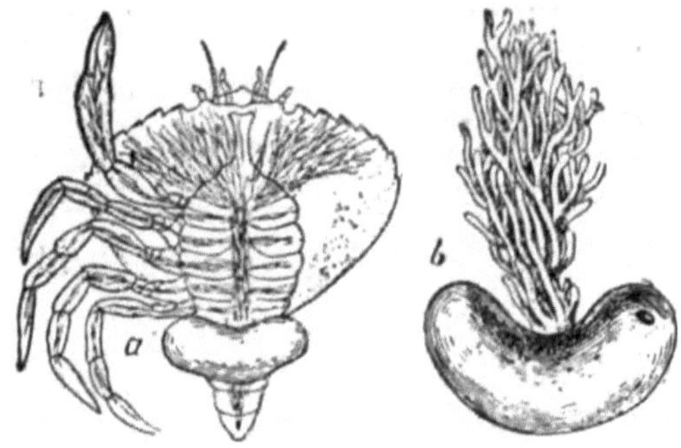

Fig. 80. *Sacculina*, a crustacean parasite of crabs. a, attached to a crab, with a kind of root that penetrates the body of its host; b. the same parasite isolated. (After Jordan and Kellogg.). This is an example of adaptation to a parasitic lifestyle and the corresponding degeneration.

OBJECTIONS TO THE THEORY OF DARWIN.

Objection	Response
The same or similar species are found in very different countries. How do we accepting that some come from others?	The subjects are separated by migration, from a common center.
The unions between different species or varieties are very different, and are almost always sterile.	This has been observed in artificial conditions and is due to the death of the embryo, inbreeding, and sexual diseases.
Many intermediate types are missing between fossil and modern species.	Cause: imperfection of paleontology, fossil destruction by natural agents and handlers. However, according to Gaudry, nowadays there are more perfect animals, some larger (whale), more intelligent (man), better flyers, runners, and swimmers.
Egyptian animals, ancient and modern, are equal, and not evolved. We do not observe that species transmute.	Inaccurate. Variation is found in all beings at all times: but this is only now being studied. No two individuals are exactly alike. In addition, the transmutation of species requires the passage of much time and natural history is a recent science: we are just beginning to study the organisms from a philosophical standpoint. (See the works cited in the introduction).
Selection is a kind of divinity that protects the most gifted beings, or an inconceivable intelligent force.	Selection is a *result* and not a *force*. The most gifted beings are retained because they protect themselves with their own resources, and all survival is mainly due to progress based on the functions of protoplasm.

Geological evidence of evolution. These are endless and we will limit ourselves to mention of the genealogy of the horse (Fig. 81) and Gaudry's two works, "The chains of the animal world," works based on facts and whose very title explains their trends and results.

Nobody can deny that the incessant progress of paleontology has almost daily robustly and almost monumentally cemented the theory of evolution.

In effect, the history of the major groups of plants and animals has been reconstructed, and it is necessary for the enemies of the theory of evolution to slowly

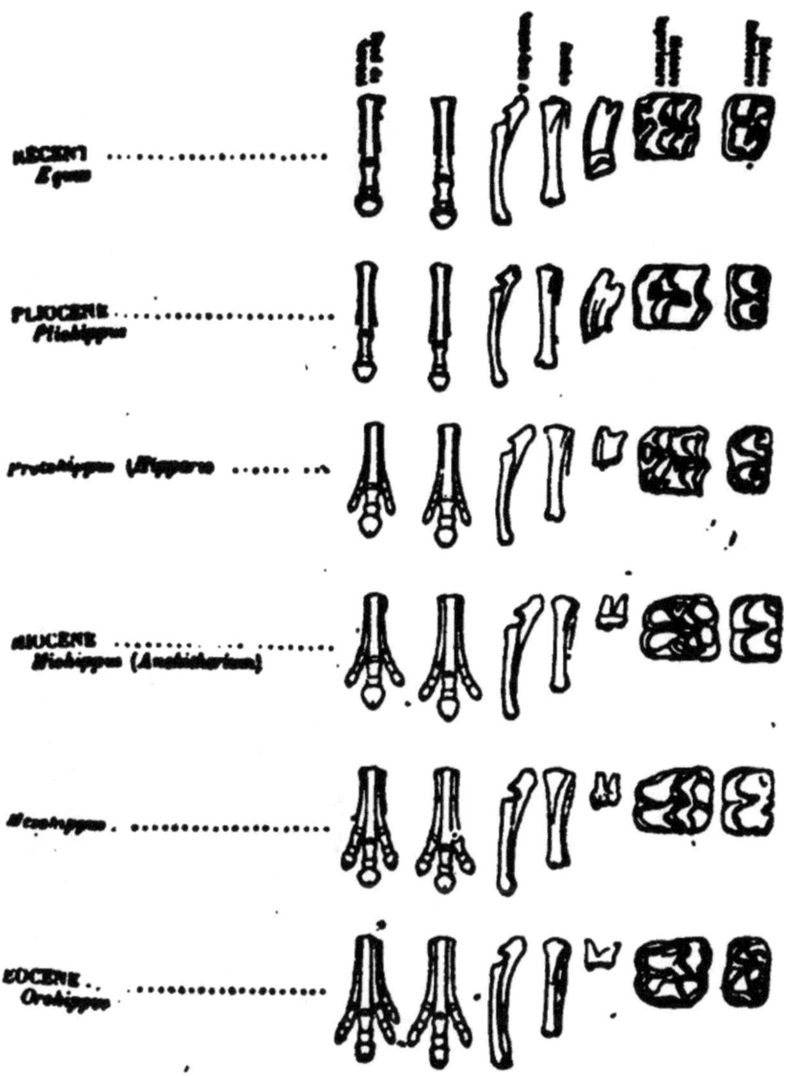

Fig. 81. The genealogy of the horse. The development of the inferior molar, the upper molar, the leg, the forearm and the foreleg. Modern horse (*Equus*), one from the Pliocene (*Pliohippus* and *Protohippus* or *Hipparion*), one from the Miocene (*Miohippus, Anchitherium, Mesohippus*) and one from the Eocene (*Orohippus*). After Wallace.

consider all that has been achieved in this research before proposing illusory or sophist difficulties.[128]

Experimental evidence. This is very rare, because the majority of naturalists and particularly the enemies of Darwinian theory, engage in trivial matters such as the

---

[128] It is impossible to summarize the findings of Cope, Marsh, Barrande, etc. in a biology textbook.

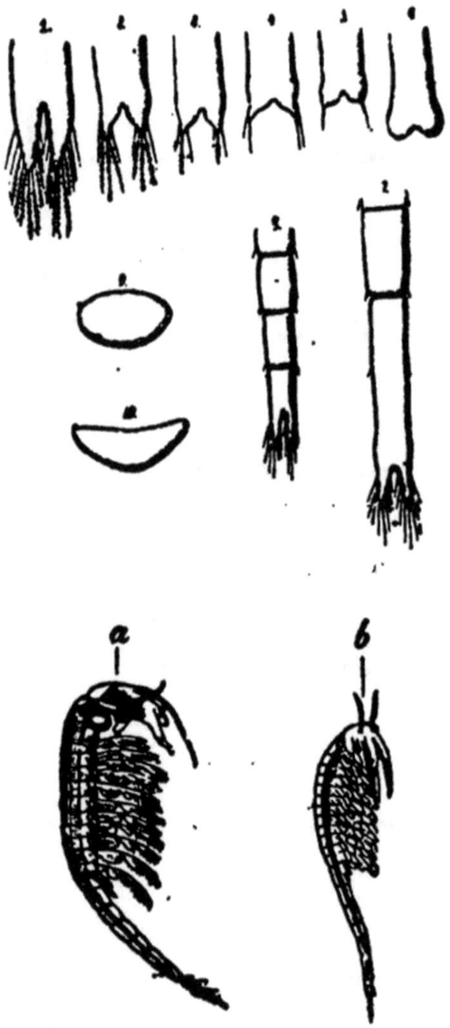

Fig. 82. Experimental tests of the theory of evolution. In the figure above: transformation of *Artemia salina*, a species of Crustacean, into *Artemia Milhausenii*, under the influence of a change in the amount of salt in water. 1. Caudal lobe of *A. salina* through 2, 3, 4 and 5 to 6, which is that of *A. Milhausenii*. 7. Post-abdomen of *A. salina*. 8. Post-abdomen of a variety from saltwater 9. Gills of *A. Milhausenii*. 10. Gills of *A. salina*. In the lower figures: a, *Branchipus stagnalis*, which is transformed into *Artemia salina, b*, under the influence of fresh water. (After Wallace.).

description of species and histological sections, rather than devoting themselves to experiments, although this is an experimental science.

However, Semper has changed the genus and the species of some crustaceans (Fig. 83) and several scholars have demonstrated profound changes of bacteria and some lower fungi, which vary in shape, color, toxicity, etc., according to the conditions under which they are cultivated.

Nascent studies in plasmogeny show that the variety of shapes and structures are due to physical and chemical causes and to details of their preparation (oleates, silicates.)

Fig. 83. The evolution of the human skull. Vertical and lateral profiles of the skull of *Pithecanthropus. P. Neanderthalensis. P. erectus. P. krapinensis.*

The evolution of man.

Man is an organism belonging to the animal kingdom, a type of vertebrate, an air breather, of the class mammalia, order of primates, which includes apes, monkeys and lemurs. However, the anatomical differences that distinguish man from apes are less than those between them and other primates. One of the characters that have been invoked to exclude man from the zoological series is the hind limbs of monkeys, which has earned these animals the admittedly false name, “quadrumanes”, but anatomy teaches us that the posterior limb of the monkey is not a hand, but a prehensile foot and the leg of the gorilla, for example, is much closer to the foot of man than to that of the orangutan. Moreover, a prehensile human foot occurs in certain savage peoples and people without arms.

Finally, the notable difference between the human brain and that of the great apes is also not an obstacle to the logical classification of this animal, as there are very signifcant analogies and eloquent transitions (children, idiots, microcephalics, madmen).

⁂

Man’s origin is uncertain, but lately very interesting fossil skeletons have been found, especially the one discovered in Java and called *Pithecanthropus erectus*: this establishes the transition between ape and primitive Neanderthal man, showing an extraordinary blend of human and simian characteristics (Fig. 84).

⁂

Science soon came to recognize the great antiquity of our species. When the famous discoveries of Boucher de Perthes were made, no one wanted to admit that they would have been contemporaneous with the extinct large mammals, and it was said that the magnificent carved flints discovered by this man of genius were simply due to shock or to chance. “Nobody has found a fossil man, thus he does not exist.”

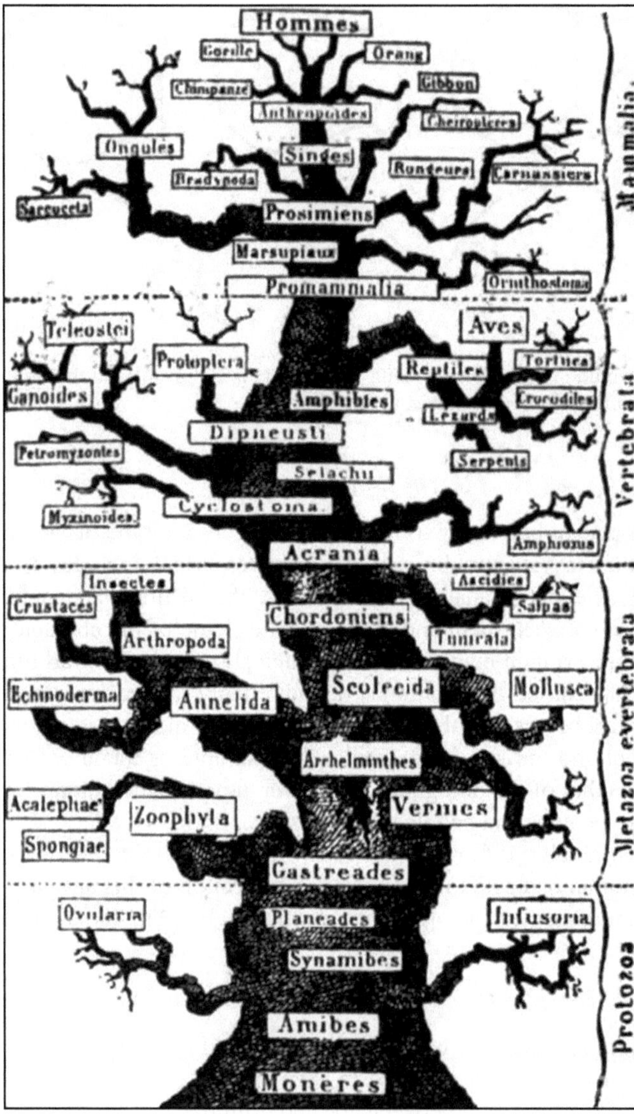

Fig. 84. *Family tree of man, according to Ernest Haeckel.* The first organisms or Protozoa were ancestors of the invertebrate Metazoans, of the vertebrates and finally, the mammals. They come from the marsupials and monotremes. The drawing shows in schematic form the evolution of the animal kingdom.

The truth is some unassailable documents are available. In the Tertiary flints have been discovered that apparently were carved by human hands, but in Quaternary formations the findings have been much more interesting and include human bones belonging to the three major races.

The oldest is extremely dolichocephalic, that is to say it has a very elongated skull, the skull is flattened laterally, has very prominent brow ridges, a small and shifty forehead, like the beard: an inferior type that perhaps occupied in Europe the space that extends from the mouth of the Rhine to the Pyrenees. In fact the remains of Neanderthal were discovered in Canstadt (Germany), and Spy (Belgium) and in some of the French dolmens.

The second race is also dolichocephalic, but the right forhead, high and wide, indicates a higher form, the face is wide and the cheeks are outgoing: this is the Cro-Magnon race, known from the remains of its industry and five skeletons. They lived in France in the age of reindeer.

In summary, the third race was clearly brachycephalic or of broad and round skull.

Boucher de Perthes demonstrated the existence of fossil man with the discovery of the famous *Moulin-Quignon* jaw, found in France. This discovery was sensational, having given rise to a new science: *paleoethnology*. The jaw was found in a deposit of fossil bones of antediluvian animals and had not suffered any removal.

***

The evolution of society, the family, marriage, religion, science, etc., occupies volumes due to the prolific pen of Lubbock and Letourneau: it would be impossible to extract them in a few pages and we will simply say that man, modern society and civilization have developed progressively by way of evolution (Fig. 85).

3rd Period: transformation.

Intra-organic struggles.

The formation of the organism by internal conditions.

The law of nutrition.

W. Roux, Delage (fig. 86) and ourselves, on a very small scale,

have stressed the impossibility of explaining the facts of evolution by *reasons of convenience* or by the *results* of struggle and selection.

There was a deep error in this, as the internal *causes* and the *forces* which form the evolving organism were confused with *needs*, the *reasons for existing*, and the *results*.

Everything in the world of organisms is or has been protoplasm and every fact of adaptation, inheritance, selection, relates primarily to the activity and changes in the protoplasm, i.e., physico-chemical causes.

Figure 87 bis, a photographic copy of a photomicrograph shows a kind of artificial ovule in the process of segmentation.

Imitation of segmentation, by means of oleic acid and neutral sodium carbonate. Upon absorbing water alkaline oleate particles dilate, dilating more the smaller the resistance that opposes their expansion.

This ovule was seen to form rapidly by dilation in water from a particle of oleic acid, within which other particles, more or less compressed by mutual pressure, took the appearance of a rope and of dorsal blastomeres, with meridians and nuclei.

Fig. 85. Yves Delage. French naturalist. Popularized the theories of Roux on the construction of the body by physico-chemical forces and the conditions of nutrition. 1900.

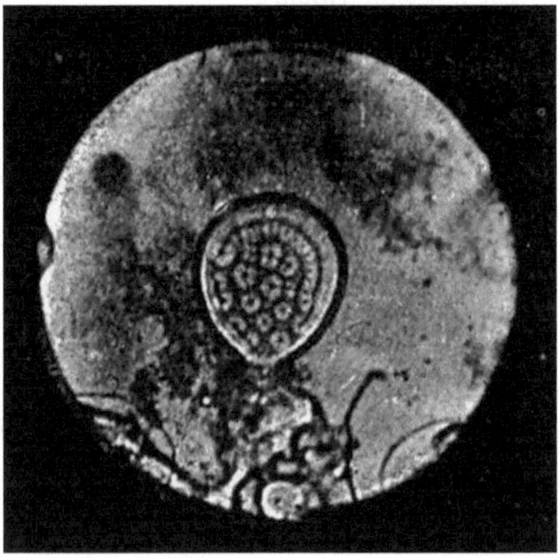

Fig. 86. bis. a photographic copy of a photomicrograph shows a kind of artificial ovule in the process of segmentation.

Fig. 87. *Chifopage monsters*, right Liao-Toun-Chen, left Liao-Sienne-Chen. These monsters came from two eggs that fused to start development under identical conditions of nutrition, they were very similar, physically and morally.

However, the natural ovule, the fetus, the adult organism, owe their structures and modifications to the internal struggle of the cells and organs for space, light, heat, and food. W. Roux has been able to explain the formation of many organs by reasons of resistance, pressure, and organic compensation (already mentioned on page 175). This science, biomechanics, is in its infancy.

The study of monstrosities teaches that nothing is fatal in humans; it is sufficient to heat a chicken egg asymmetrically to form a deformed chicken with a hypertrophied heart and various more or less interesting abnormalities, the kind that the common man attributes to curses and spells and that the old physiologists explained in a metaphysical manner.

But there's more: a necessary generalization supported by all positive biological knowledge has made it clear that *all* of the phenomena in the *organism are related to the mechanism of nutrition (the law of nutrition)*. By nutrition we mean the relationship between the organism or the cell and the internal and external environment. Thus, two eggs, developed in the same ovary and therefore with very similar nutritional conditions, will form twins, almost identical physically and morally, and even chipofage monsters (Fig. 87).

Thus the theory of evolution has progressed in its turn consists of two sets of issues:

1. Effects: evolution, the struggle for existence, variation, selection (Darwinism).

2. Causes: Physico-Chemical, nutrition.[129]

[129] To doubt these conclusions would be to confess great ignorance of the most authoritative modern work. See the work of Roux and his "Archives of biomechanics."

Synoptic table and summary of the theory of evolution.

COMPARISON OF THE MAIN EXPLANATIONS.			The three periods of the theory of evolution.	Evolutionary principles and precursors.	Laws that support the theory of evolution.	*1st Law: Reproduction*
Metaphysical.	Cuvier.	Lamarck and Darwin.				
The Creator formed beings in a single providential act.	There were several creations and general disasters.	Beings have gradually formed from a single-celled organism through slow variation and selection of the most advantageous in the struggle for existence.	1st, creation, 1480 (B.C.) to 1832. 2nd, demonstration, 1830 to 1863. 3rd, transformation, 1885 to 1904.	*Moses*.-Genesis. *Linneus*.-Classification. Order. *Natura non facitsaltum.* *E. Darwin.* Importance transformations species. *0ken*.-animal origin of man. Primitive colloidal substance. *Lamarck*.-Unity of Evolution. Action of the environment. Slow, physico-chemical causes. Spontaneous generation. (1809.) *Geoffroy*. Unity of plan. Evolution. Correlation of the organs. (1830.) *Goethe*.-Metamorphosis of the organs. (1832.) *Lyell*.-Slow geological causes. (1830.) *Spencer*.-Unity of Evolution. Medium. (1858.) *C. Darwin*.- (1858), and *Wallace* (1858.) Selection. Struggle for life. Evolution.-*Huxley*.-Id. *Haeckel*.-Id. Monism or the unitary conception of the universe. (1863-1 904.) *Delage* and *Roux*.-Construction of the body by internal conditions. Law of nutrition.	1st Law of reproduction. 2nd Law of organic correlation or compensation. 3rd Law of inheritance. 4th Law of Malthus. 5th Law of the persistence of simple shapes. 6th Law of the struggle for life. 7th Law of natural selection.	Fertility and gestation period are related to the struggle for life, the smallest species persist thanks to their prodigious fertility. Domestic species are generally more fertile than wild ones. Organisms transmit life to their offspring with varied, but not identical characteristics.

Synoptic table and summary of the theory of evolution.

2nd Law: Organic correlation or compensation.	3rd Law: inheritance.	4th Law: of Malthus.	5th Law: the persistence of simple shapes.	6th Law: the struggle for life.	7th Law: selection.
The modifications of a given organ determine those of others. Cave animals lose their sight but their sense of touch is exquisite. When a plant bears many flowers it produces few leaves and vice versa. The tail and other organs of the human embryo atrophy, due to development of the brain and elsewhere. This law is very important for the explanation of individual development or ontogeny.	*Acquired modifications:*hereditary if they change the general nutrition. *Hereditary diseases:*epilepsy, insanity, heart problems, weakness and predispositions. *Atavism:* the return to the primitive state (wild horses). *Consanguinity:*almost always fatal. Inheritance is due to reasons of nutrition and conserves individual characteristics for some time.	Geometric progression of species and arithmetic progression of individuals. This form is defective, it is too precise. In general, the lack or scarcity of food inevitable when a species multiplies greatly.	Because they are less complex and more adaptable, they adapt better and vary less. Panchronic species occur across time. (Successive spontaneous generations?)	All beings struggling for life, with the weather, epidemics, food, very fertile omnivores are more persistent (rats). *Mutual relations:*clover, beetles, mice and cats. *Symbiosis:* nitrifying bacteria in legumes. Cross-pollination by insects. Means of defense. Passive or active. Mimicry. Association. Mechanical, physical, and chemical means. The struggle for life, pain and suffering, are essential for progress. Darwin demonstrated and made known the importance of this law.	Natural or artificial. Certain varieties or forms persist better than others, because they are more advantageous or more desirable to man. *Consequences:* incessant divergence of characteristics, extinction of species (walrus), which do not reappear and are intermediate between others, in geological strata, or predecessor of the current in an isolated location. Nature does not advance by leaps and bounds, but by slow variation and selection. There is unity of plan (bat wing and fish fin.) Organic progress of nutrition and its apparatuses and resources. *Facts explained:* geographical distribution, incessant dissemination through the natural or physiological barriers, by active or passive, periodic or accidental migration. Persistence of rudimentary organs (blind gut. Selection is a result, not a *force*.

Synoptic table and summary of the theory of evolution (Continued)

APPENDIX.

VARIOUS TESTS OF THE THEORYEVOLUTION.

Objections to the theory.	*Response.*	*Geological evidence.*	*Experimental evidence.*	*Evolution of Man.*	*Period of transformation. The formation of the body by internal conditions.*
Same species in different countries. Hybrid sterility. Missing links. Species that have not changed, in Egypt. Selection would be an intelligent force.	Their origin was the same, but they have separated by migration. Under artificial conditions. Diseases. Consanguinity. Imperfection of paleontology. Constant conditions. Small variations. It takes time for others. It is a *result* of variations and the progress of nutrition.	Chaining of plants and animals in successive geological ages. Genealogy of the horse, the antlers of deer, the higher molluscs.	Transformation of *Artemia salina* into *Artemia Milhausenii* and *Branchipus stagnalis*. Transformation of bacteria and some lower fungi. Artificial pseudo-organisms from oleates and silicates.	*Pithecanthropus erectus. P. neardenthalensis.* Fossil races of Cro-Magnon, Moulin-Quignon, etc. Evidence drawn from prehistory and history.	The struggle for space and nutrition in general is verified actively inside the body and cells and tissues which are better equipped to better defend themselves against phagocytosis, etc. persist. There are no internal constructive forces. Everything is related to nutrition. (Twins are very similar, like the conditions under which they form.)

Synoptic table and summary of the theory of evolution (Continued).

APPENDIX.						
CLASSIFICATIONS	VARIOUS TESTS OF THE THEORY OF EVOLUTION.					
All are conventional and arbitrary. In nature there are only individuals in the process of evolution. The species was the group of individuals that are fertilized and reproduce among themselves, but there are fertile hybrids, varieties seem like species, and species that seem like varieties. Neither the characters of analogy and adaptation nor those of more physiological importance can be used to make perfect classifications, because everything is essentially unique. Classifiers are waning naturalists and will soon disappear, being supplanted by the physical chemists.	*Ontogeny. Metamorphosis. Colonies.* Individual development *(Ontogeny)* is equal in its broad outlines, to that of the family *(phylogeny.)*Human embryos with gills, tail, etc. The larva is a free embryo and its metamorphoses are progressive. Higher beings are colonies of Protozoa or Protophyes.	*Polymorphism and variation.* The leaves of plants vary according to the incidence of light. All beings vary greatly, as do their organs, and the supposed fixity of species is a myth. Many variations in size (dwarfism), color (polychrome), geographical, seasonal, etc. are known.	*Adaptation. Convergence.* Beings and organs adapt to needs and different conditions. (Cosmopolitan species, senses and hands of man). Similar needs produce similar modifications (marsupials, carnivores, rodents, herbivores, pithecoids.) Flying fish, insects, reptiles, birds and flying shellfish.	*Division of labor.* Is the criterion of improvement. The vibrating lashes of the infusoria are used for locomotion, secretion, respiration, functions that have distinct organs in higher animals. In bees and ants there are different kinds of individuals, neuters, males, females, with different obligations. Division of the sexes.	*Parasites. Degeneration.* Species that are subtract themselves from the struggle for life, degenerate and lose the organs of locomotion and some related ones. However, organisms are progressing as a whole and it would be impossible for all of them to become sedentary and degenerated parasite. (Human society today.)	*Mendel's laws.* White varieties when crossed with gray Descendantsare gray. This color or character is dominant, the other, latent or recessive. The germ-cells of the hybrids are just one of the characters of the generators Soon it will be possible to predict the outcome of crosses, in all details.

BOOK THREE.

SUMMARY OF CONCEPTS IN BIOLOGY. -CONCLUSIONS.

1. Whither man on Earth?
2. Whither matter in the Universe?

Biology is the science of life, of the material phenomena of the organism, which in the past and present have been or are caused by physical-chemical forces known.

Life, a prolonged agony, is not the struggle between the alleged life force and physical-chemical forces, but the decay, the collective consecutive organic creations and destructions, the osmotic activity of the protoplasm, the physical basis of organized beings, this kind of foam, emulsion or network of infinitely small alveoli, which entrap elements from the environment, transforming them incessantly.

And like all the structures of the organisms are or have been protoplasm, all biology is reduced to the obstinate consideration of protoplasm, whose manifestations range from dormancy or of dried organisms; the oscillating state, the active state, the aestival state, sleep, varieties recognized because of the delay or acceleration of osmotic flows, flows that only death paralyzes irremediably (and even the corpse has its biology, according to Delage).

But this science is not exclusively speculative, it forms the basis of agriculture, medicine, sociology, and why not say bravely, of human happiness.

Biology is, in effect, a new science, which aims to explain things, not to contemplate them, considering them in motion, not at rest, not like paralyzed parts in museum display cases, like petrified mummies; Biology is reduced, as all science, to a sub-field of general mechanics or etherology, since the universe is a unitary fact and a great truth, because everything created is a great republic, headed by a democrat, movement and everything sensible is reduced to mass and motion. In this way the unity, the law of great things, is presented, evident when comparing the Macrocosmos, the Microcosmos and the Mesocosmos (simple grouping invented by the human mind to facilitate study); upon investigating the mutual transformation of forces, the probable existence of a fundamental element of all matter, whether the protile or the electrical atom or electron, and finally, upon continuing detailed microscopic, chemical and physical examination, of the essential substance of organisms or protoplasm, processing apparatus of forces like human or celestial machines, apparatus of ultra-microscopic gears, where life exists without consuming its own energy self, subordinate to the universal principles of thermochemistry and the general characteristics of movement and circulation which also manifest themselves on a planet like Earth, pseudo-living organism, a type of cell of the ether, heated by the sun and revolving around the sun, with its active organized and inorganic inhabitants, with its polar condenser and equatorial boiler, with its immense continents slowly migrating over the solidified crust, that rises and falls or whose secular tide or brittle matrix is continuously eroded, destroyed, rebuilt by sludge agents of telluric physiology, volcanos, air, water, the sea, Life.

***

The Universe is a unity forever mutable in form, forever unchanging at its base. Living bodies or organizations can be made with a special substance and consist mainly of salts, carbon, oxygen, hydrogen and nitrogen. Recent studies have shown that crystals are formed by the condensation of a kind of nucleated protoplasm, called petroplasm (von Schroen), that is to say that have a similar origin to that of the cell and real germs of crystals are known, which are grown as a species of microbes and produce similar crystals and can heal and regenerate. Moreover, the best imitation of microscopic creatures are made with inorganic reagents, such as silicic acid. In sum, there is no fundamental difference between organic and inorganic chemistry and it is only necessary to find out some details of the formation of albuminous substances, which are produced in the plant under the influence of light and minerals, air, water and salts.

Beings can be considered hypothetically as colloidal minerals. Even albumins can be crystallized, so that almost all the principles of our organism extracted from a corpse by chemical means may acquire a particular geometric shape.

It is impossible that organic materials come from the same organic substances, and it is believed rather that the living being is composed of a kind of salt (silicate, phosphate in combination with nucleonic acid with nucleins) of colloidal or soapy consistency with very fine alveolar structure, in which the elements of air and water are condensed, forming various organic bodies. But this hypothesis has not been proven yet.

However, the studies of J. Gaube (du Gers)[130] confirm the importance of minerals "that are indispensable for the manifestations of life. (There are biodynamic metals such as iron). Ferments, hidratases and oxidases owe their activity to various principle minerals." These works of generalization and coordination should be better known as rudimentary logic compels us to agree on the now very generalized idea, that inorganic living beings existed before organic ones, when they were unable to withstand the very high temperatures of early geological periods. Spontaneous generation probably took place under conditions very similar to modern ones and even today, many materials, albumins, fats and other tertiary or quaternary ones are forming in plants and animals, and in man himself. As silicates and colloidal silica abound everywhere and tend to be organized and combined with other bodies, which retain avidly, one wonders if they are in fact the desired inorganic basis of life. Although they do not strongly resemble life, at least they look wonderful due to their physical and chemical characteristics.

***

The unity, the law of great things, as we said at the beginning of this short summary, is at the bottom of all protoplasms, in such a way that the apparent differences

[130] Cours de Minéralogie Biologique. 4 Series. 4 Volumes. 1899–1908.

between animals and plants are superficial, and every living thing breathes, feeds, reproduces, grows and dies. However, selection, the struggle for life, has led to a progressive evolution or development of all the species that, without losing

1 Cours de Mineralogie Biologique. 4 series. 4 volumes. 1899–1903.

their deep analogies, diverge more every day, acquiring organs and systems well suited for the accumulation and processing of food, that is to say energy. (See the summary of the doctrine of evolution.)

***

Whither man on Earth?

Where will all the trappings of man on earth, rocks and plants, all animals, the organic, the inorganic, the planet, the satellite, the sun, matter evolves in the Universe. Because of the beliefs, superstitions and endless desire that all men have to survive their tired and monotonous earthly existence, they have built this absurd theory of the impossible isolation of our species, which should only be associated with divinity, today, finally, man understands that he is only one of many living beings, which have unified characters that are descended from others, for the preservation of the fittest in the struggle for survival.

***

Living pseudo-beings and pseudo-organized structures have been prepared in laboratories, with reagents that are not mysterious or divine, like oil and potassium carbonate, tannin and gelatin, oleic acid and alkalis, ferrocyanates, phosphates, and silicates. The latter, the basis of mineralogy, seem to be also the basis of physiology, in artificial amoebas and fungi, already so perfect that they can be mistaken for diseased or dormant organisms, with a kind of indifferent life. In this way the analogies between inanimate and animate matter are now becoming so suggestive, that the mind is confused, like one of those surprised nocturnal birds surprised by unexpected and vivid light, stunned, hesitating before making the final and definitive conclusion: *that there is no gap between living and crystallized bodies*. The idea that the cube of salt or *aluminum* prism are formed by the condensation of a petroplasma and can be broken, scarred, or assimilate, as a Protococcus or as a *rose petal*, is sufficient for the true scientist to accepts, openly and boldly, the unified mechanistic theory of nature.

***

Whither man on Earth? Where do all the trappings of man on earth, rocks and plants, animals all, the inorganic and organic, planet, satellite, suns, go as matter evolves in the Universe.

Thus concludes modern philosophy and this proclaims the universal destination, the endless migration of the infinitely large and infinitely small, the exodus of the sands, the race of the powders of the Cosmos, in one of whose atoms, the world still hears the mournful voice of a vertebrate and unguiculate placental species, man asks this question tirelessly “where am I and where am I going?” like a miserable slowly dying insect attached to the frame of a locomotive, a monster of steel and fire, running full speed, climbing glaciers and cliffs, with frightening speed, into the mystery of the night.

ADDITIONS AND ERRATA.

The abundance of silicic acid in various organs and organic products according to Hugo Schulz.[131]

Meat	0.10 to 0.54 per cent by ash.
Aorta	0.28
Tendons of veal or beef	0.48
Vitreous body	0.15
Spleen	0.16 to 0.45
Skin	0.14
Dura madre	0.33
Pus	0.05
Ovarian cyst	0.03
Embrionic gelatin	1.69
Gluten	1.56

A. Hilger found silicic acid in the ashes of the Falusias, salps and sea squirts, Forscher, in the cuticle of the holothurians, 0.57 percent; Weber, egg white, 0.28 percent; Poleck in idem. 0.49 to 2.04 and 0.55 to 1.40 percent in the yolk, Henneberg, in the blood of the hen, 0.96; Gorup-Besanez, in granivorous bird feathers, 40 percent; in carnivores, 27 percent, the same in the insectivores; in fish-eaters, 10.5 percent.

Ox blood, 1.11, percent, Weber.- Ox-Blood, 2.81, Stölzel.-cow milk, 0.06 to 0.09, Weber.- Ox bile, 0.36, Rose. – Beef, 2.07, Stölzel.-Beef, 0.81, Staffel. -Deer hair, 8.1, Gorup-Besanez.- sheep hair, 8.3, Gorup Besanez.- goat hair, 9.4, Gorup-Besanez. – porpoise hair, 9.4, Gorup-Besanez.-Ox hair, 10.8, Gorup-Besanez.-Rabbit hair, 11.8, Gorup-Besanez.-Dog hair, 12.5, Gorup-Besanez.-Horse hair, 14.6, Gorup-Besanez.-Negro human hair, 6.61, Baudrimont. -White human hair, 12.30, Baudrimont.-Dark hair, 30.66, Baudrimont.-blond hair, Baudrimont 30.71. -Red hair, 42.46, Baudrimont.- woman milk, traces, Wildenstein.-Brain, 0.42, Breed.-Liver (56 year old) 0.27, Oidtmann.-Liver (senescence) 0, 11. 0idmann.-Liver newborn, from syphilitic parents, 0.17, 0idmann. Blood of a rheumatic, 0.53, Witting.

[131] Ueber den Kieselsäuregehalt menschlicher und thierischer Gewebe. Archiv. f. d. Gesammte Physiologie. Pflüger, Bonn 1901, Bd. 84, pp. 67 to 100. -In another article (Zur Physiologie und Pharmakodynamik der Kieselsäure. Deutsche Med. Wochenschr. 1903, no. 38) the author says that silicic acid has important physiological and medicinal properties, producing in excess, cerebral symptoms, rash, hair loss, great development of the nails, digestive disturbances, stiff joints and bone pain.

Dr. Srhulz, says, "the form and state of combination of silicic acid in connective tissues should be studied, then do not reveal their way of life when treated by solvents, and must also consider the possibility of the silicon atom may exist in the place of carbon in the albumin of these tissues, according to the discovery of Ladenburg. He concludes by ensuring that silicic acid is always found where there is connective tissue, it being very interesting that it is found at the same time in superior animals and siliceous sponges, as well as in the swim bladder of fish. Its therapeutics applications should be very interesting."

There are, finally, large chemical analogies between silicon and carbon[132] and an alcohol is also known in which some of the carbon is replaced by the silicon.[133]

This requires the conclusion that our imitations of protoplasm, prepared with silicic acid and silicates, tend to be confused, by deep and evocative characters, with the natural living model, it being necessary to note that Arthur Müller, in his classification of colloidal bodies, including silicic acid in the second group, where it contains organic colloids, which coagulate changing chemical state, while the colloids formed by fine particles in suspension, like colloidal silver, has no osmostic power and coagulates by electro-capillary causes.[134] And if the elements of air and water are combined under the influence of light, this combination could only be verified in a mineral matrix, since nuclein and general protoplasm cannot exist without phosphorus, silicon and other inorganic elements.

[132] Analogies du silicium avec le carbone. Friedel et Ladenburg. C. R. t. LXIV.

[133] C. R. t. LXI. Friedel y Ladenburg.

[134] Müller, A. Ueber die Klassifikation der Kolloide. Bioch. Centr. Bd. II., p. 56.

Plasmogeny

Prof. A. L. HERRERA
Director of Biological Investigations, Department of Agriculture, México, D. F.

Plasmogeny is the name proposed by the author for a new science, which includes a study of the life-like behavior of artefacts, as well as all researches and theories which serve to throw light on the origin of protoplasm (from *protos,* first; and *plasma,* form).

From a study of artefacts and their behavior and "life," reasonable inference may be drawn as to the factors at work in living cells, tissues, and organs.

Plasmogeny makes use of the data and conclusions of many other sciences. From chemistry it appropriates knowledge about proteins, nucleoproteins, carbohydrates, lipoids, and all organic and inorganic substances found in living matter. From physics and physical chemistry, it takes what is known of colloids and colloidal behavior, including structure, surface tension, diffusion, osmosis and osmotic pressure, adsorption, electro-capillarity; electric charge, etc. From biology and its subordinate sciences, such as cytology and physiology, it draws data concerning the behavior and function of living matter (protoplasm, nucleus, cell, tissue, etc.). Hence the following divisions:[1]

Morphology
- Cytogeny, or imitation of cells.
- Histogeny, or imitation of tissues.
- Organogeny, or imitation of organs.

Physiogeny, or imitation of functions.

Chcmiogeny, or imitation of organic molecules concerned in life processes, e.g. Polypeptids, proteins, lipins, cellulose, Chlorophyll, and such substances as vitamins, hormones, bios, etc.

[1] For more general application of the philosophical principl es of Plasmogeny, see publications by A. L. Herrera, e.g. "La Vie Universelle." *Bull. Assoc. Int. Biocosmique.* Chatenay-Malabry. Paris. May. 1926. No. 1. p. 1–16—"Una nueva ciencia. La Plasmogenia." Maucci, ed. Barcelona. 1926, pp. 1–446, illustrated.

H. J. Cleaves et al., *Herrera's 'Plasmogenia' and Other Collected Works,*

DOI 10.1007/978-1-4939-0736-6_3,

The illuminating work of Pasteur showed that ordinarily there is no such thing as "spontaneous generation," the production of organisms without parents or antecedents from inorganic or decomposing organic matter, in culture media.[2]

This led to the aphorism: *omne vivum ex ovo,* which has since dominated biological science, although Tyndall, for example, said that he saw in matter the potentiality and possibility of all life. In fact, on reflection, it is evident that protoplasm and life must have been the results of natural energies and forces in the primitive environment, which might well have been different from present conditions. We cannot, therefore, accept this aphorism, but look forward to the possibility of discovering the conditions under which some form of primitive life will arise.[3] Though the goal may still be in the far ofï future, some steps have been made. Many of the substances of living cells have been produced by photosynthesis.

Résumé of Some Results Obtained

Mr. Maynard Shipley says[4] "The recent researches of M. Daniel Berthelot show, among other efïects, that the ultraviolet rays cause decomposition phenomena in most organic Compounds. To the decomposition phenomena of *electrolysis* we must now add *phoiolysis,* which is found to have an even wider range than electrolysis, not well limited, as it is, to certain classes of bodies in Solution or in a molten state. One very remarkable effect produced by photolysis is that substances exposed to the

[2] We, to-day, can hardly appreciate the efïort that Pasteur had to put forth to establish this view; but in those days "scientific" formulas were extant, which told, e.g. "How to produce three mice" by the use of a mess calculated to attract denizens of the laboratory. *J. A.*

[3] The fact that no one has yet demonstrated the spontaneous or the laboratory productioni of a living thing, is no argument against the view that spontaneous generation has occurred, and may, indeed, be even a common occurrence. The trouble has always been that, instead of searching for. the origin of life in its ultramicroscopic rootlets, attention has been concentrated upon living forms tremendously advanced in the scale of evolution. Life did not begin bv the sudden appearance of an elephant, a mouse, a flea, or any of the other forms of life which entered Noah's Ark. Since Loewenhoek's time the microscope has revealed hosts of living things (bionts), both plants and animals, unknown to the writers of the Biblical record: but even these are quite complicated, as any book on Protozoology will show. The law of probability is strongly against chance grouping of the relatively enormous numbers of molecules needed to make up an amoeba or a bacterium, but much more in favor of the chance formation of a living molecule or molecular group, as described in the first paper in this volume.

Obviously, the artefacts shown by Leduc and Herrera, illustrate not living thirigs but rather many of the physico-chemical concomitants of life. They enable us to form an idea as to the development of some of the curious and even weird forms we find in plants and in animals, many of which would be scouted as figments of the imagination, were it not for the fact that they exist. The sources of life lie in the submicroscopic field, below the level of these artefacts, and are demonstrated indirectly. e.g. by genetics, just as the chemist indirectly demonstrates atoms and molecules by their experimental effects. J. A.

[4] "Man's Debt to the Sun." Little Blue Book No. 808. Edited by E. Haldeman-Julius Girard, Kansas, U. S. A., pp. 1–64. E. Slosson, "Creative Chemistry," New York, 1920.

action of the ultraviolet rays will give ofi gases in a manner resembling what occurs in fermentation." Moore and Webster, Baly, Heilbronn and Hudson and many others studied the syntheses of formaldehyde and of nitrogenous products from inorganic subtances under the influence of ultraviolet light. Spoehr, Porter and Rampsperger[5] and the author have not attained success in these experiments. I query if there is not required a condenser, a structure probably formed by inorganic colloids; this is also the idea of Ciamician.[6] Silica and the silicates produce, as inorganic colloids, an infinite series of imitations of organic phenomena, and they exist everywhere in nature and in organic matter and cells (*see* Herrera[7] and the bibliography). Silica absorbs minute quantities of radium, which acts on cells and life. Organic photosynthetic substances, e.g. formaldehyde, also form imitations of cells and mitotic figures.[8]

Plasmogeny, originated by Noliet, who discovered osmosis in 1748, was a child about 1885, with the work of Butschli and Quincke. To-day it is an adult in possession of its full strength and faculties. Who knows when it will reach its objective, which is the synthesis of living matter?[9]

Organic Colloids

Vibert varnish dropped in water. Imitation of cells and infusoria in movement by changes of surface tension and vapors of amyl alcohol.

Tannin in gelatin and water, at 70°C. Imitation of cells and microscopie beings. The calcium chloride of the gelatin is precipitated by the silica of the tannin. Firm structures. *See Bulletin de la Société Zoologique de France,* Jan., 1899, p. 20; 1902, p. 144, fig.

Oleic, capric and caprylic acids and alkalis and alkaline carbonates, in water. A great variety of myelinoid forms, structures, movements, ciliary motion, striking resemblance to Protozoa and Protophyta. Cells, nucleus, spiremes, membranes,

---

[5] "The Action of Ultraviolet Light upon Carbon Dioxide and Water," J. Am Chem Soc 1925 p. 79.'

[6] "Gaceta Médica Catalana," 1921, p. 24. Revue Scientifique, Paris, 1921, n. 6, p. 140, and n. 20, p. 603.

[7] Herrera, R. Accad. Lincei. Roma, June 15, 1924, pp. 510–512. Figs.

[8] Researches of A. L. Herrera, Mexico, 1889–1926, 36 volumes published, 5400 experiments "Una nueva ciencia. La Plasmogenia." Maucci, ed. Barcelona, 1926, pp. 1–446, about 400 engravings and plates.

[9] At the 1926 Meeting of the British Association for the Advancement of Science, Miss F S Semmens reported experiments showing hydrolysis by light polarized by colloidal particles. She and others point out that this peculiar action of polarized light may be the germ of truth underlying many popular superstitions, e.g., that food exposed to moonlight readily decomposes. The belief in potency of the moon also crops out in "moon-struck," "lunatic" and in certain Chinese ideographs. Miss Semmens' work has been confirmed and extended by Prof. H. I. Macht of Johns Hopkins University, reported at Richmond, Va., meeting of American Chemical Society, 1927; see Science for April 15, 1927 (No. 1685). *J. A.*

direct division, films, palpitations, tubes, neurones, contractile threads, spirals, trees, amoeba in movement, plasmodia of Myxomycetes, ovules and embryos, etc. All the figures soluble after several days.

Colors and cross in polarized light. They are "liquid crystals." *See:* Lehmann-Herrera. *Mémoires de la Société Scientifique Antonio Alzate,* Mexico. Vol. XII, p. 242, figs.

Oleic acid in the surface of lime water. Macroscopic amoeboid movement.Palpitations, like a heart.' Osmotic currents. *Bull. Soc. Zoologique France.* Jan. 24, 1899, p. 21.

Oleic acid dissolved in carbon disulfide or tetrachloride, in ammoniacal water. Macroscopic amoeboid movement, vacuoles, conjugation and direct division. The same with oleic acid dissolved in chloroform. *La Médicine Scientifique de Quesneville.* Paris. December, 1911, No. 132, p. 179-182. Oil and sulfur chloride and hydrochloric acid, in alkaline water. Amoeboid movements.

Olive oil or rosin dissolved in gasoline. French olive oil (F. Bétus & Fils) 50, gasoline 100, or rosin 6, gasoline 240, and drops of a Solution of caustic soda (14%) stained with rhodamine (1 gr.).

Remarkable imitations of amoeba and infusoria.—"Colpoids" (Fig. 1), artificial imperfect beings attacking and sucking themselves (Fig. 2). These osmotic cells show internal currents and a great variety of macroscopic and microscopie structures and movements, chiefly determined by changes of surface tension and the propulsive action of osmotic currents. Many functions: conjugation, palpitation, deformation, contractile vacuoles, pseudopodia, direct division and multiplication, growth, increase of surface and volume. On adding gum arabic the movements persist one hour. Adding lactose: *autophagy,* the arms of some bifurcated colpoids suck each other. On adding oleic acid, the membrane is thinnest and the pseudopodia are largest. *Phagocytosis* of carbon impregnated with acetic acid. These experiments are far-reaching and throw some light on muscular contraction and indicate some of the conditions of primary osmotic cells and membranes. Colpoids are the best imitations of living infusoria, and present the most remarkable results of Plasmogeny up to 1926. Upon desiccation or rest, colpoids show structures and nucleus in division. The same Solution of soda dropped from a height of 20 centimeters on gasoline and olive oil, presents a marvelous imitation of Brownian movement, each granule being activated by osmotic currents and not by the kinetic energy of the liquid. This 'is a proof of the author's biological theory of Brownian movement (La *Semana Médica de Buenos Aires.* Feb. 6, 1913, p. 305-308. May 28, 1914, p. 1227-1238, figs.; Fig. 3)

Theory of colpoids. In the photomicrographs we see the osmotic currents and the currents between two or more colpoids. Doubtless the cause of the sucking activity, resembling a vital function, is the continuous change of density of soda Solution contained within each colpoid. This change is produced by the saponification by the alkali of the oil dissolved in gasoline. The density of the Solution drops continuously. Then this density is not the same in two colpoids or in the different parts of a colpoid; and active currents of diffusion, by difference of density, are

Fig. 1.—Artificial "amoebas" (magnified).

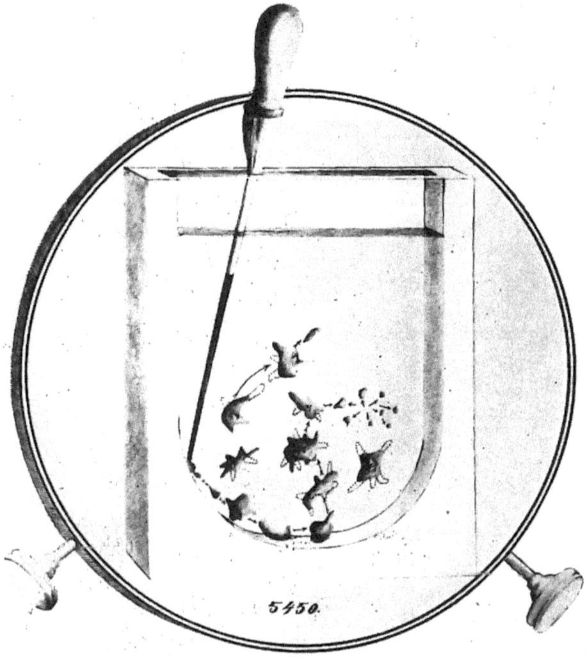

Fig. 2.—Colpoids, illustrating conflict and parasitism (slightly magnified).

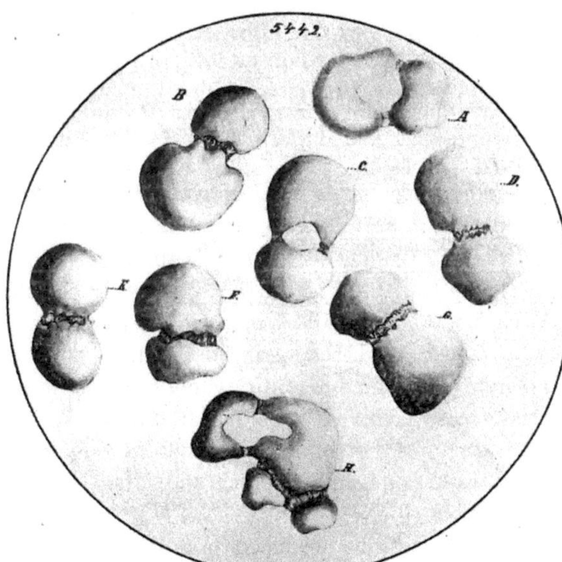

Fig. 3.—Colpoids, showing nucleus and division 24 hours after production (slightly magnified).

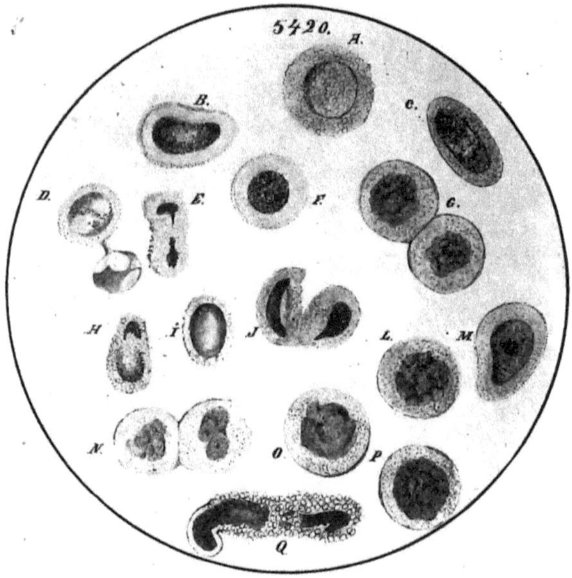

soon established between the artefacts and its membranes tili they attain the same osmotic pressure and the same density.

This theory is also, I believe, far-reaching in biology. The pseudoplants of Traube and Leduc are also the result of two different densities of the exterior and the interior liquid, as probably are a great many biological phenomena. *See,* for colpoids, *Compt. rend.,* Feb. 15, 1926. *La Côte d'Azur Médical.* Toulon, France, June, 1926, No. 118-119, fig. *Buil. Muséum d'Histoire Naturelle,* Paris, 1926, No. 4, pp. 218-220, lig. *Buil. VAssoc. Internat. Biocosmique, La Vie Universelle,* Vol. 1, pp. 13-16, 17-20, fig.[10]

Collodion diffused in a paper soaked with linseed oil. Cells and nucleus with chromosomes. The collodion should be stained with methylene blue. Macroscopic.

Hexagonal tissue of collodion. Formula: alcohol (90%) 250 cc.; ether 240 cc.; glycerin 8.40 cc.; ultramarine blue 6 grams; ivory black 1 gram; aluminium in fine powder 5.520 grams; gold bronze in fine powder 6 gr. Cellular whirls of Dauzère, Cartaud and Bénard. Hexagonal tissue, nucleus, embryos, fine structure of protoplasm, etc. They are due to the evaporation and formation of concentration nuclei, each nucleus becoming hexagonal by pressure. If evaporation is prevented, there are no structures. For Imitation of mitosis, this liquid is poured on a glass plate, a drop of water is put in the center, and strong aspiration is applied by an air pump and two tubes (Fig. 4).

Splendid asters are produced, mitotic figures, with a spindle and chromosomes. *(See* "Una nueva ciencia, etc.," p. 292-293, figs.)

[10] The instability of colpoid protoplasm is like that of natural protoplasm. The death of colpoids is a result of instability.

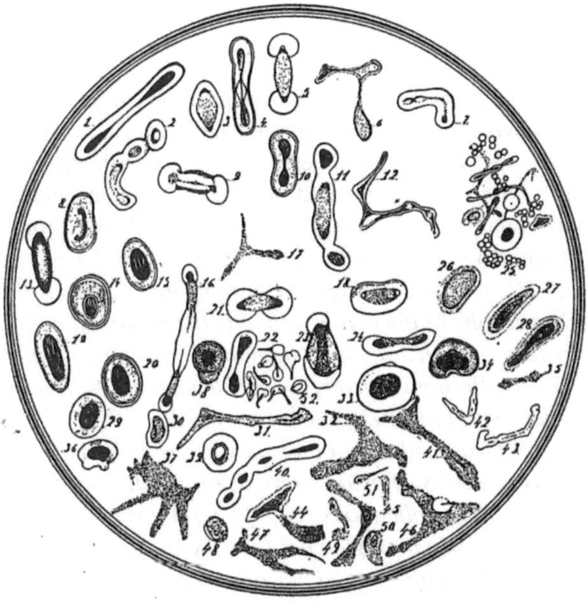

Fig. 4.—Some imitations of cells made with calcium fluorosilicate; fixed, stained, and mounted in Canada balsam. (From *R. Accad. Lincei.)*

Formol cvaporated in a desiccator. Cellular and mitotic figures, imperfect crystallisation. This experiment is eviclence favoring the theory of the photosynthetic origin of life proposed by Herrera in 1905, in *Revista Chilena de Historia Natural,* Aug., 1904, pp. 235-237. *See: Rend. accad. sei. Lincei,* April, 1925, p. 364, and Jan., 1925, p. 5. Baudisch.' *Semana Médiea*Buenos Aires, Dec. 21, 1922, p. 1319.

Gelatin and glycerin. Injection of air or gas: polyhedrical cells, tissue, imitation of insects' Compound eyes, multiple images. *See* "Botánicä" by Herrera. Herrero Hermanos, México, editor. 1924, p. 404. Lima Scientific International Congress, 1924. Herrera, "Una nueva ciencia. La Plasmogenia," p. 263.

Albumin and alkalis. Imitation of fibers. *Bidl. Soc. Zoologique France,* Jan. 24, 1902, p. 23. *Mémoires de la Société scientifique Antonio Alzate.* Mexico. Vol. XI, pp. 29-31. *Rev. Sei.,* Jan. 10, 1902.

Albumin and phosphoric acid. Movements and structures due to impurities (soluble silicates). *Bull. Soc. Zoologique France,* April 8, 1902, p. 159-164, figs. Structures of gelatin and metaphosphoric acid. *Ibid.,* June 24, 1902, p. 177-201. *Mém. Soc. Alzate,* 1902, p. 133-136, figs.

Egg albumin and acetic and metaphosphoric acids. Silicic Structures. *Bull. Soc. Zoologique France,* July 22, 1902, p. 201-202, figs.

Rosin and gasoline (20: 2.50) and drops of soda Solution (14%). Microscopical amoeba, retractile pseudopodia. Soluble in gasoline. This is the best imitation of amoeba produced in a laboratory.

Inorganic Colloids

*Sodium silicates, colloidal silica, organic Compounds of silica and fluoro-silicates mixed with salts, alcohol, sulfuric ether, acids, for*mol, oil, etc. *Silica is found every where.* Cells, animal and vegetable tissues in thin sections, slowly incinerated, show silica. Also in organic matter (tannin, proteins nucleins, crystallised chemically pure sugar, etc.). Probably silica is a constituent of organisms. See Herrera, "Sur la presence de la silice dans *les coupes histologiques incinérés."* Compt. rend., *Feb. 16,1925, p. 538. "Biología y Plasmogenia," Mexico, 1924, p. 245-286, with the full bibliography of the matter. Silica is the universal colloid and shows morphological diastasic properties. See Zsigmondy, "Coloidequímica" Calpe, Madrid, 1925, p. 267-297. Perhaps silica is the substratum of life. It adsorbs radio-active substances (Ebler and Fellner,* Z. anorg. Chem., *73, 1-30 (1912). The best imitations of mitosis are produced with silica and substances rieh in silica,* e.g., *collodion and oils. An extensive series of models produced by silica and fluorosilicates may be found in the author's publications.* Compt. rend., *May 19, 1919, p. 388, June 28, 1920, p. 00. Importanza biologica dei fluosilicati.* Rend. accad. sei. Lincei, *Jan. 7, 1923, p. 42-44. SuH'imitazioni dei piccoli detagli dei Microsporidi col fluosilicato di calcio.* Rend. accad. sei. Lincei, *1925, p. 639-643, figs (Fig. 5).*

Silica exists in diastases (Matignon, *Compt. rend.,* Feb; 1, 1924) and shows diastasic activity (A. Mary and others). Bachmann has observed the production of pseudo-animated forms by evaporation of a silicic gel impregnated with benzene, *Z. anorg. Chem.,* 73, 165 (1911). *See* Herrera, "Notions de Biologie et Plasmogénie," W. Junk, Berlin, 1906. The author believes that there are no imitations of living beings without silica. Cells of calcium fluorosilicate can be washed, fixed, stainecl and mounted in Canada baisam like natural cells, and the author has sent hundreds of these prepara- tions to European laboratories. Silica is an oxidation catalyser [Shibata and Kaneko, /. *Chem. Soc. Japan,* 45, 155, 78 (1924)], has therapeutic and nutritive properties and has been employed as manure by Oberlin, Lemmermann and many others. There are a great number of organogels and organic Compounds of silica that may be the basis of a new chemical world (Fig. 6 and Fig. 7).

Mercury. The author has published many works on the Beilstein amoeba (*Bol. Dirección Estudios Biológicos,* 1, 211-254). Hydrosomes and mercurisomes. Pseudo-artificial beings in movement. Irised films. In- fluence of silica. Microscopical results. Imitation of Infusoria. (*Ibid.,* 2. 40, fig.)

Mercury and aluminium. Conical growths in moist air, etc. (*Bol. Dirección Estudios Biológicos,* México, 2, 40, fig.).[11]

Hydrochloric acid and sodium Silicate. Imitation of fungi, *Puccinia, Exoascus,* Diatomaceaë, *Pseudopeziza,* etc. Microscopie, hard. Infiltrations of the acid into the Silicate. Cause: imperfect crystallisation of sodium dhloride in silica. Stained and mounted in Canada balsam. *Mém. Soc. Sei. Antonio Alzate,* 26, 1-49, figs. *Arch. Plasmologie Générale,* Bruxelles. Lamertin, 1, Fase. I, p. 21, pl.

[11] See also paper by H. Wislicenus. Vol. I, this series. *J. A.*

Fig. 5.—"Spores" made with calcium fluorosilicate.

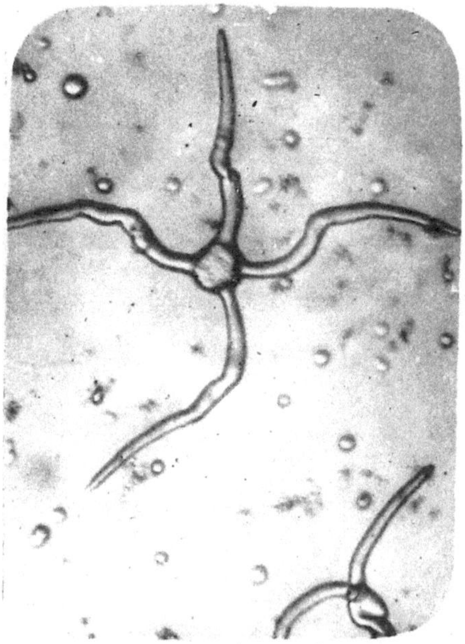

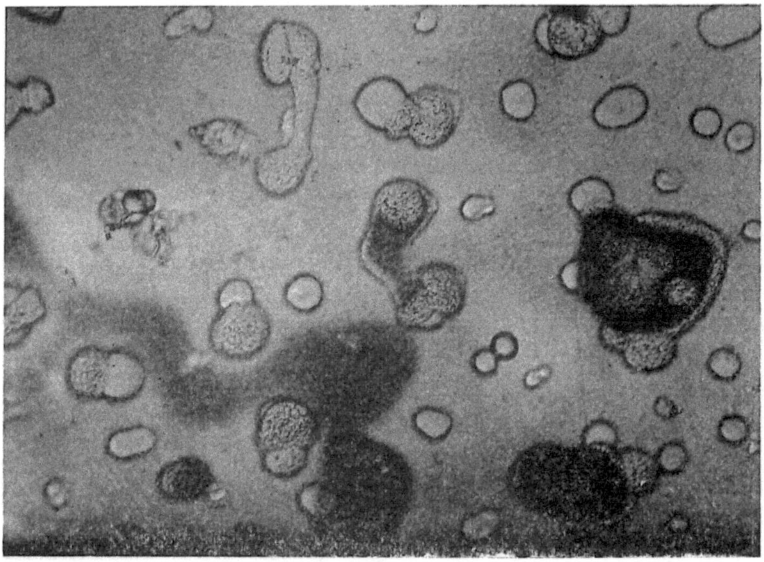

Fig. 6.— Artefacts (X 153), produced by dropping solid $CaCl_2$ into a mixture of egg white and sodium Silicate, containing some KF.HF.

Fig. 7.—Artefact produced by slow diffusion of $CaCl_2$ into a Solution of sodium Silicate containing a mixture of various salts, including KF.HF.

Potassium chloride and silico-carbonate of potassium and calcium chloride. Crystallisation and, some time after, cellular forms. Amoeba. *See* Castel- lanos, "La Plasmogenia," Habana, 1921, p. 142.

Potassium silico-carbonate and calcium chloride. Reversible production of crystals and cells. "Protobios," 1917. Castellanos, *ibid.,* p. 141.

Potassium silico-carbonate in a porcelain candle, placed in a Solution of calcium chloride. Nucleated cells in division. *J. Lab. Clin. Med.,* 4, 479.

Infltrations of calcium chloride in a Solution of Silicate and fluoride of potassium, in capillary compressed films, at 90°C. Complete cells with nucleus and filaments. Mounted and stained in slides. Many years of work and many publications. *Compt. rend., loc. cit.* "Biologia y Plasmogenia." México, p. 266-305. (*See* Figs. 3 and 4.)

Calcium carbonate in silica and silicates, and in colloidal silica. Spheroliths, cells, amoeba. Albumin in Harting experiments *is not necessary.*

Mémoires Soc. Alzate, 26, 277-279, figs. "Comité de la Alianza Científica Universal," México, I, p. 19-21. The silica of reagents and albumin explains the forms and results of Rainey, Harting, Burke, Dubois, Kuckuck, etc. In all imitations of life silica exists and must be sought for,

Silicates and manganese salts. Artificial oxidases. *Mém. Soc. Alzate,* 29, 331. Pseudo-plants of Traube prepared with manganese Silicate decompose actively H_2O_2.

Colloidal silica and carbonates or sodium and potassium. Evaporation of the Solution. Cells and imitation of human embryos. Cause: imperfect crystallisation of sodium carbonate in hygroscopic potassium carbonate, water and colloidal silica. *Tcrapeútica Moderna,* México, 31, No. 2, p. 6. *See* also: Mary, "Dictionnaire de Biologie physiciste" (Paris), p. 67-69, figs.

VARIOUS IMITATIONS

Colloidal membrane pressure or hymenopiesis. Bol. Direeción Estudios Biológicos, México, I, 228-229, figs. 231-232, 235-236. Collodion dried on mercury; expulsion of metallic drops.

Imitation of plants, buds, leafs, glands, bowels, brains of animals and of man by plastic solid matters injected in skulls or compressed in linen or sereens. Bol. Direeción Estudios Biológicos, I, 341-366, figs. Castellanos,. La Plasmogenia, p. 33-39, figs (Fig. 8).

Alcohol dropped in stained suicate films. Imitation of cells, nucleus, chromosomes, neurones, etc. *Bol. Plasmogenia,* Habana, I, 93-94, figs. *Homo,* Bruxelles, I, p. 80-82, fig. *UÉcho Mexique,* Paris, p. 11-12. *Rend. accad. sei. Lincei,* May 6, 1923, p. 508-510, figs.

Vibrations of lycopods in a mcmbranc, forms of cells and infusoria. Bol. Dirección Estudios Biológicos. México, 2, 45-62, figs.

Imitation of rhythmic coloration in insects, birds, cells, tissues, muscles, ova, glands, nucleus, storch, etc., by means of calcium fluorosilicate. Explained by Sedimentation. "Semana Médica," Buenos Aires, September 21, 1922, p. 615-621, figs.

Imitation of Brownian movement. Drops of caustic soda (14% Solution) in gasoline and olive oil (50%) from a height of 60-80 centimeters. Biological theory of the Brownian movement. "La Vie Universelle," *Bull. Assoc. Internat. Biocosmique,* I, no. 1, p. 13: *Mém. Soc. Sei. Antonio Alzate,* 32, 209-211; *La Terapeutica Moderna,* México, Vol. 23, no. 9, p. 65-68, fig. *Semana Médica* de Buenos Aires, Feb. 6, 1913, p. 305-308; May 28, 1924, p. 1227-1238, figs. These researches explain the granulations of colloids by a biological theory, and the general property of Brownian movement in colloidal matter. The experiment with drops of Solution of soda is striking and should be repeated in every laboratory.

FUTURE WORK. SYNTHESIS OF THE PROTOPLASM WITH AMINO-ACIDS

According to modern ideas protoplasm consists mainly of proteins, nucleo- proteins, lipoids, water and salts. The production of a protoplasm endowed with Organisation is a formidable task. The work of d'Herelle[12] and others on *bacteriophage*

[12] See paper by F. d'Herelle, this volume. *J. A.*

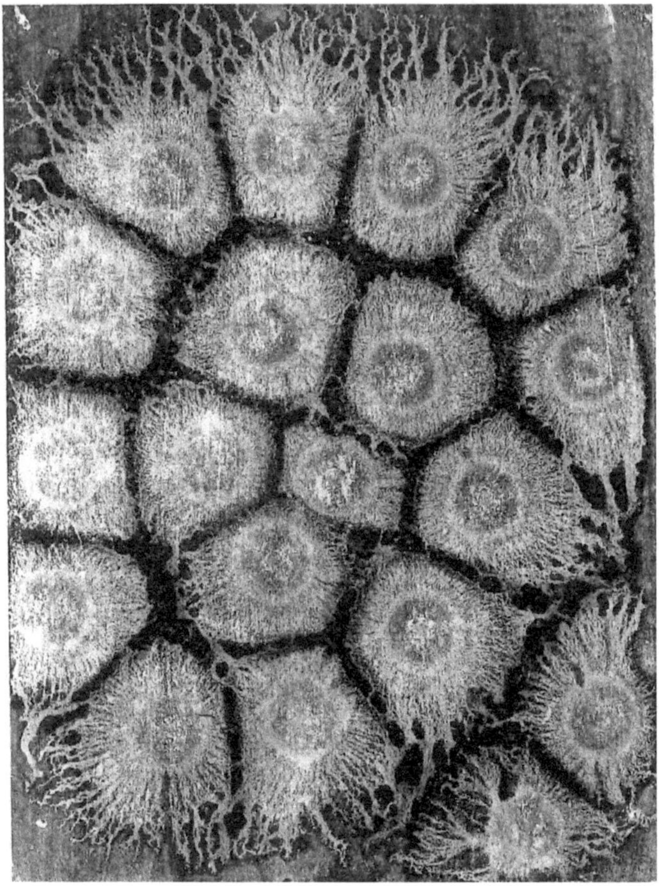

Fig. 8.—Imitation of Neurones:. Alcohol dropped on sodium Silicate films containing lamp-black, etc.

indicates how great is the complexity even of a bacterium. We here approach the border-line between "animate" and "inanimate matter."

Photosynthesis may have produced initial substances by reduction of CO2, and also amino-acids according to the work of Moore, Webster, Baly, D. Berthelot and Gaudechon, Baudisch, Allen and Church, etc. The author has lately obtained "colpoids" with drops of soda (14%) and glycine or amino- acetic acid in gasoline and olive oil, with some lability.

The early work of Butler-Burke and Dubois on radiobes indicates that as a silica gel can adsorb the most minute quantities of radio-active substances (Ebler ancl Fellner) perhaps we shall prepare the Protobies, or initial beings, by means of a mineral nutritive Solution, uranium salts or radium salts, and silicates and fluorosilicates. Radium may have furnished the primeval energy of the life, before photosynthesis determined by Chlorophyll, which could not precede life.[13]

"It is known that the rays from radium have the power to stimulate all forms of life, even to the extent of speeding up to the growth of plants and of making dormant plants burst into bud.... It is because of its property of emitting negative electrons (beta rays), that potassium is a necessary constituent of all living matter. It may, however, be replaced under certain conditions, by other radio-active substances. Certain of the rays decompose ammonia, and water under their infiuence is subjected to electrolysis, yielding oxygen and hydrogen."[14]

---

[13] Maynard Shipley, "Wonders of Radium." Little Blue Book No. 1000. Haldeman-Julius, Girard, Kansas.

[14] *See* S. C. Lind, D. C. Bardwell and J. H. Perry, "The Chemical Action of Gaseous Ions Produced by Alpha Particles. VII. Unsaturated Carbon Compounds." *J. Am. Chem. Soc.*, 48, 1556–75, 2335 (1926). *See also* S. C. Lind, "The Chemical Effects of Alpha Particles and Electrons," The Chemical Catalog Company, New York, 1926, pp. 1–180; Carleton Ellis and Alfred A. Wells, "Ultraviolet Rays," The Chemical Catalog Company, New Vork, 1926, illustrated.

PLASMOGENY, A NEW SCIENCE OF THE ORIGIN OF LIFE

PREFACE

The Church has always hampered the progress of Mankind so as to retain its hold over awareness, which is the basis of secular exploitation of the people.

Thus, it defends the tenets of the Bible zealously - attacking, persecuting and suffocating any attempt to explain the Nature of Science.

We live, without having asked for it, enduring the harshness of an ever-painful existence, the end of which is always tragic.

What is life? What is the origin of life? What is the underlying cause? Why does it manifest itself?

This has been the great enigma of all ages.

Religion states: "God created us and the origin of life is exclusively supernatural and divine, comprising body and soul; the former- a fragile shell - disappears after death, while the spirit returns to its Creator and is eternal."

Science says: "Life is movement in the infinite, universal, without separation between matter, referred to as inert or dead, and organized beings; it is due to the physical and chemical forces, and every organism: man, grass or insect, is nothing more than a product of chemicals, an agglomeration of material substances, without a soul, which fully expires upon death, to end up as water and carbonic acid: the gas that makes the water in siphons bubble."

Naturally, accepting the dogma of the Church, people are ruled by the clergy, who are associated with monarchy and militarism, mitigated in the Republic, inexistent now in Russia, where there is no religion and where, nevertheless, they live without enduring the wrath of God, who does not exist.

The mission of Science is therefore deeply libertarian and Science alone can bring relative happiness to nations, based on culture, free-thinking - with proven facts and theories instead of superstition, without domination or tyranny, without monarchs or priests.

But after a terrible struggle between Science and fanaticism, there are still ulcers, errors, lies - postulates of which Science is aware and destroys or cures slowly, fighting with great strength the millions of devotees, who from their academies and universities, insinuate a veto, or "frontier" to human knowledge.

H. J. Cleaves et al., *Herrera's 'Plasmogenia' and Other Collected Works,*
DOI 10.1007/978-1-4939-0736-6_4,

Such has been the enigma of the origin of life, and therefore the Church, worshipper of Pasteur, the fanatical genius, alas, foams at the mouth with rage at Plasmogeny, the new science of the origin of life, and seeks to destroy it, which can never be done, because all Human science advances swiftly towards the natural explanation of our origin.

In accordance with the above, in this booklet I present a summary of the new doctrine so that people can henceforth banish from their thoughts the dogma of the creation of living beings. Life has come about under the influence of blind natural forces, like a mountain, a world, a hurricane or the undulation and murmuring of the waves.

To live is to perform a physical and chemical function. Nothing more.

A. L. HERRERA

Mexico, 12th October, 1932.

CHAPTER ONE

Outline

DEFINITION AND SCOPE OF PLASMOGENY.

The word “Plasmogeny” is composed of two Greek words: *plasma*, meaning molded form or protoplasm, and *geny* meaning production.

Protoplasm is a mixture of substances and water, like a gel or thick paste that fills the cell, the building block of the body, like a little box or unit, or living independently as in the case of the amoeba. (See definition in section 6.)

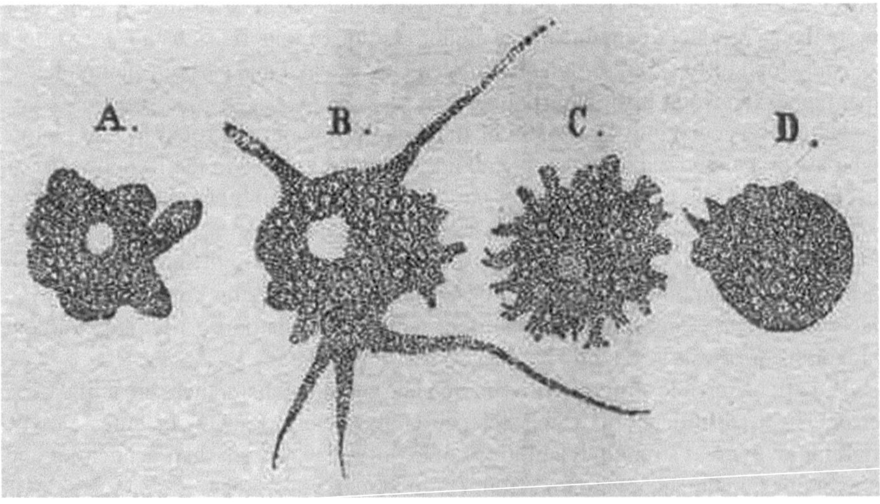

Fig. 1 The challenge of Plasmogeny: to artificially produce this microscopic being, which creeps along, changing its body shape, in stagnant water. The cell is filled with protoplasm

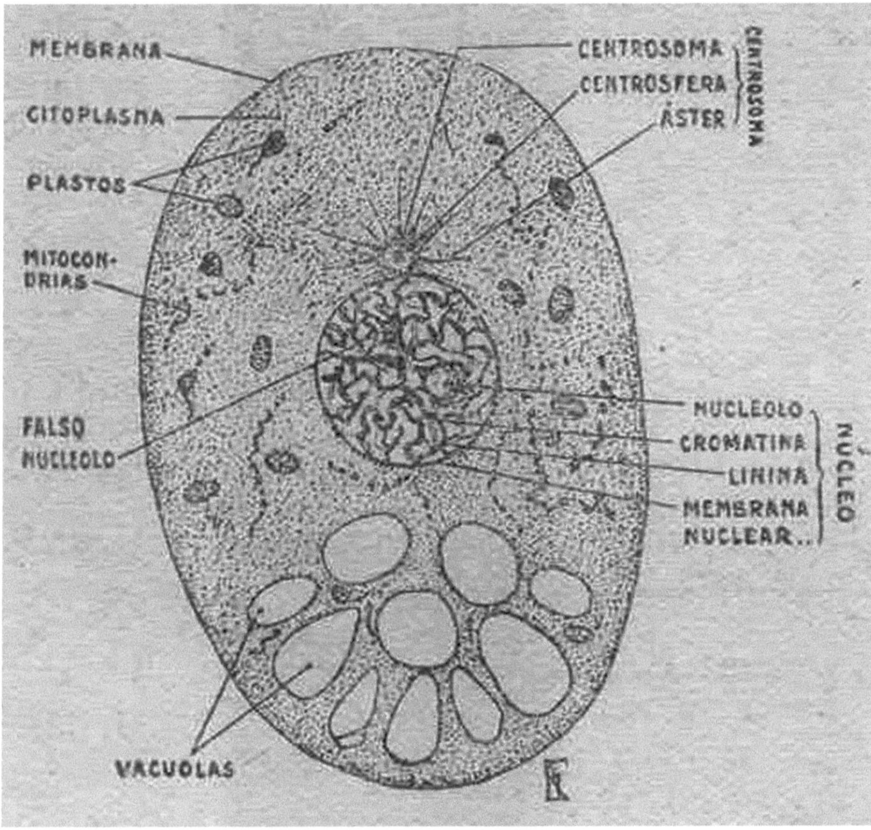

Fig. 2 Also called cytoplasm.- Diagram of an animal cell, greatly magnified. (To avoid repetition, note that whenever cells are represented, one is to assume they are visualised at high magnification).

Protoplasm has been called the physical foundation of life, but it would be preferable to say the physicochemical foundation of life, because not only do physical properties, like density, underpin its form, but also its chemical composition.

Plasmogeny is an experimental science, which endeavours to study the origin of protoplasm, just as Cosmogony, for example, investigates the origin of the universe, and Pathogenesis, the origin of disease.

In fact, everything is alive, Nature is unique; everything moves, externally or internally, due to a primitive substance or plasma.

Space is filled with a kind of electrical fog, which has been called ether, and as Plasmogeny is all-encompassing in its meditation and examination, the above definition can be generalized, stating that Plasmogeny is the science of the *universal protoplasm* and *its origin.*

A comparison in layman's terms clarifies these ideas: We take a lump of clay or putty and shape it as we please, into a plant leaf, a hand or even a person. Clay represents protoplasm.

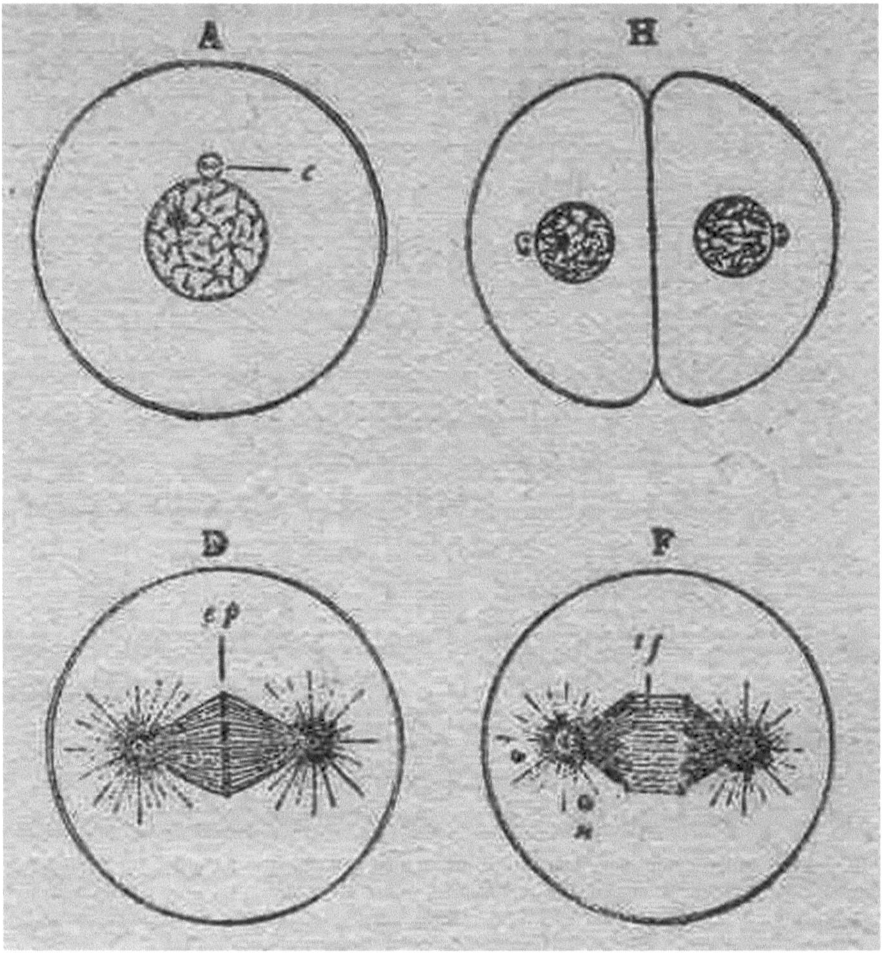

Fig. 3 Cell Division (according to Cendrero).

PLASMOGENY IS A SCIENCE

Generally speaking, Science refers to certain knowledge of things in terms of their principles and causes. Science is a methodically acquired and organised doctrine or body of instruction, which represents a particular branch of human knowledge.

Plasmogeny is, therefore, a science, with a well-defined and important objective, following logical methods based on observation and experimentation.

Mechanics, for example, is the science of movement and the forces that can produce it, although it is considered part of another broader science: Physics.

Natural History or Biology is the science of life, comprising two major sciences: Zoology, or animal science, and Botany or plant science, in addition to Mineralogy also known as mineral science.

One must understand that Plasmogeny encompasses a huge number of works by *scholars from the world over and from all epochs*, it is not limited to a single scientist, in which case it would just be a personal opinion, questionable and possibly mistaken. On the contrary, it has been studied by internationally renowned geniuses like Berthelot, Fischer, and Claude Bernard, with its indirect founders being Jean Lamarck, the French naturalist, predecessor of Darwin, and author of the theory of evolution, Claude Bernard, the French scholar who founded the branch of science called Physiology and Marcellin Pierre Berthelot, the famous French chemist and founder of chemical synthesis also known as the artificial manufacture of chemical substances, previously proffered as evidence of a life force underpinning living beings, and believed to be the only force able to produce them.

Thus is Plasmogeny conceived and set forth by the author of this booklet, and not as a set of his own and very modest works.

Indeed, it is derived from human knowledge as a whole; knowledge which is woven together in the most intricate way, allowing, nonetheless, divisions and branches of science to organize and facilitate the study thereof.

In addition to the three sages mentioned above, many others, from every country and era, have contributed to building this new science, among whom we should mention: Runge, 1865; Nollet, 1748; Buffon 1788; Dutrochet, 1824; Rose, 1837; Link, 1839; Böttger, 1865; Traube, 1866; Harting, 1871; Monnier and Vogt, 1882; García Díaz, 1885; Graham, 1862; Bütschli, 1885; Leduc, 1901; Lehmann, Gaubert, Mary, Delfino, Renaudet, Bambaren, Lillie, Fischer, Torres, Buscalioni, Lecha-Marzo and many others cited in specialized works. The author of this publication has dedicated thirty-six years to Plasmogeny, proposing it as a new science in 1903.

As of October 1932, he has performed over 8,200 (eight thousand two hundred) experiments, which have been meticulously recorded in his laboratory notebooks.

As Plasmogeny encompasses many branches of knowledge more or less intimately, its documentation is vast - with colloidal or gelatinous substances alone (the basis of protoplasm) occupying countless books and pamphlets and a specialised German Gazette, which has published numerous volumes so far and is unfinished. The treatise on these substances, published by Mr. Jerome Alexander[1], comprises four volumes, totalling 3,392 pages and 202 articles written by specialists from round the globe. One of the first chapters deals with Plasmogeny.

There is a whole host of books, journals, institutes, laboratories and specialists devoted to various topics encompassed by Plasmogeny, and this profusion of material demands the formation of a new science that will organize and present it in an orderly and rational way. The list of works and pamphlets on Plasmogeny can be found in the publications of Mr. Victor Delfino, published in Buenos Aires, chiefly in two newspapers: *La Medicina Argentina* y *La Semana Médica*.

The late Prof. Jules Félix of Brussels founded the Institute of Plasmogeny, which was sadly destroyed upon the death of its founder in the midst of dark intrigue involving self-serving relatives and, I believe, the Catholic clergy that swarm in Belgium.

---

[1] *Colloid Chemistry*, New York. *The Chemical Catalog Co.*, 1926-1932.

One area of Plasmogeny, dealing with liquid crystals, which are life-like, occupies many works and there are purpose-built microscopes and accessories and an arsenal of around a hundred substances produced by these crystals, perhaps five hundred.

Finally, organic chemistry dealing with carbon-based compounds, synthetic chemistry relating to artificial compounds, and photosynthesis or energy production by means of light, are fruitful branches of Science, which are richly and exceptionally documented.

PLASMOGENY IS A NEW SCIENCE.

Different sciences are distinguished by their object of study. As soon as a different object can be demonstrated, one which can be studied and known, a particular science can be recognized to exist. And what could be more important than discerning the origin of our own life, the origin of protoplasm.

Plasmogeny is not Biology, which is considered a "diphthong" of Zoology and Botany, as it embraces life in general and is based on the properties of material, which is improperly called dead, like carbon, which in special combinations and forms makes up our body and that of all beings.

Plasmogeny is a new science, which has advanced using elements of modern experimental research, precision apparatus, microscopes and ultramicroscopes. The former enable one to see very small objects, beings and their parts, with the light reflected by a mirror placed below the objective, while the ultramicroscope allows one to see much smaller objects illuminated from the sides. An infinite number of apparatus and study equipment assists specialists in their work, *e.g.* X-rays have greatly contributed to our knowledge of the structure of crystals, which is closely related to that of protoplasm, as we shall see later.

DIVISIONS OF PLASMOGENY.

It is naturally divided into the *abstract* and the *concrete*. The former involves the examination of concepts relating to universal life and generalizations derived thereof, appreciable in its relationship with those arising from other sciences, for example, the idea of universal life, which encompasses the stars and their origin. Because it is most interesting that experiments lead us to unite Nature, unexpectedly, so often the imitations of cells mimic astronomical forms.

The latter, or concrete Plasmogeny, involves Laboratory research and leaves the task of appropriately comparing, interpreting and generalizing results to the abstract facet.

- PLASMOGENY
  - Concrete
    - Pure
      - Morphogenesis
        - *Cytogeny,* simulation of cells
        - *Histogeny,* simulation of tissues
        - *Physiogeny,* simulation of the functions
        - *Chemogeny,* simulation of organic molecules and proteins and nucleins, which make up the protoplasm.
    - Applied
      - to Medicine
      - to Agriculture
  - Abstract
    - Pure
    - Applied
      - Sociology
      - Physiology

THE SCIENCES PRIOR TO PLASMOGENY DO NOT SHARE A FUNDAMENTAL GOAL NOR ARE THEY UNITED BY A SCIENTIFIC PRINCIPLE. PLASMOGENY UNITES THEM AND PROVIDES A KEY OBJECTIVE.

It seems incredible that no-one has ever developed, boldly and accurately, the scientific goal, which is and only is the *study of universal life*. Nothing could be grander, generalizing our own existence to dead matter, atoms, space, stars, nebulae and other celestial bodies, the whole microscopic and macroscopic ensemble.

Fanaticism has taken great care to separate Science and its branches, divide the enemy to defeat it, isolate it from Literature and Philosophy, from Art and Poetry, which should, on the contrary, as Renán wished, form a whole, brimming with beauty and allure, thus reaching the heart of the people and uniting the gifts and desires of thought.

A PLASMOGENIC DEFINITION OF LIFE.

Life is the physicochemical activity of protoplasm, a colloidal system of natural origin, derived from carbon, oxygen, nitrogen, hydrogen, sulphur, phosphorus and other known elements, and natural forces.

A colloidal system is a mass of gelatinous substances, like egg-white, gum, oil and fats in general, bound to mineral matter, undergoing constant evolution and change, unstable due to infighting between the crystallisable or crystalloid portions, and the non crystallisable ones or colloids, which are exposed to crystallization.

Experiment.- To explain and clarify these concepts the reader can do the following experiment: Dissolve a little gum Arabic powder in water and salt, then spread the liquid on a glass slide and leave it to dry slowly in the shade.

The next day, under the microscope one can see imperfect cubic crystals, of salt, assuming spherical forms, like eggs, like stars, because the rubber colloid does not let the salt molecules and atoms cluster in geometric forms. According to recent studies, the cell and the protoplasm are formed by similar imperfect crystallizations. By adding sodium hydroxide to stearic acid dissolved in petrol and then leaving it to evaporate, imperfect crystals are also produced, having the appearance of natural cells (Fig. 4).

PLASMOGENY OBEYS LOGIC. NEW SCIENCTIFIC METHODS AND EXAMPLES OF THEIR APPLICATION.

It recognizes the value of Method, and employs both its procedures: analysis and synthesis. These procedures are the same as for other sciences: hypothesize, compare, analyze, induce and deduce, etc.

LAWS OF ABSTRACT PLASMOGENY.

Fundamental unit.- The universe is one great whole; no star, force or being is made according to special laws or with special elements or particular purpose, and everything comes down to movement/motion. "There is nothing in the intelligence that has not first been in the senses."

In a series of publications, the author of this book attempts to show that the Universal Drama is represented by all that exists and rules in all catastrophe, strife, extermination, shock, destruction, pain, while the pleasures, which optimistic philosophy exaggerates, are negligible. Perhaps what exists, comes from an ultracosmic

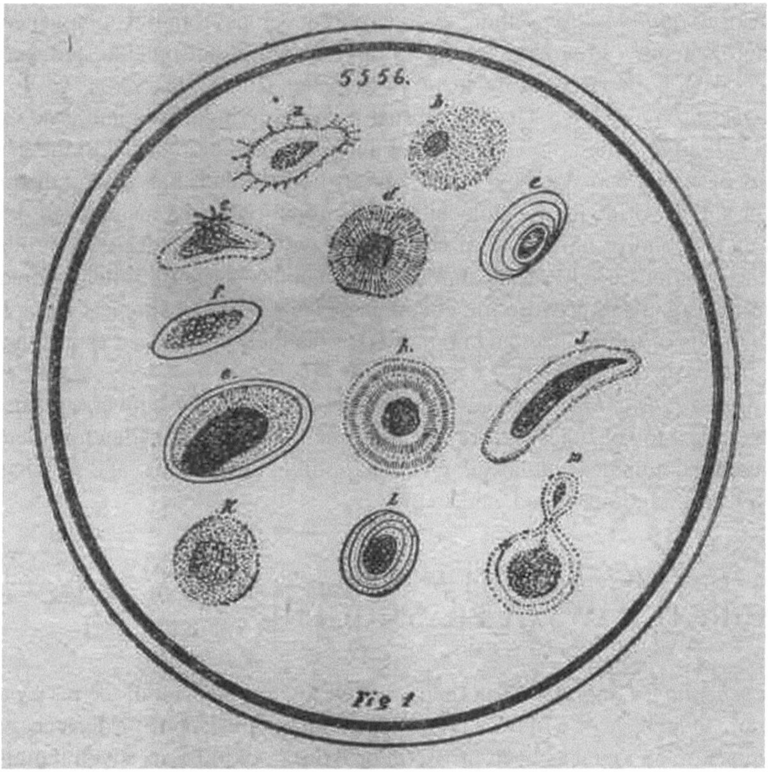

Fig. 4 Sodium stearate (Herrera). The latest theory states that protoplasm consists of a very complex substance, of atomic and molecular structure, based on proteins, due to the amino acid bonds, similar to the product obtained by condensing ammonium thiocyanate with formalin. (See the last part: "Current Works")

catastrophe, and everything is sinking. According to the latest research, the Universe is expanding and extending, the nebulae are receding at very great speeds, like smoke scattered before the wind[2].

Rather than the joyous and gracious creation of a good-natured and joyful God, we must acknowledge universal pain, the great tragedy of matter and of life, confirmed by Astronomy, Geology or the history of the Earth and its inhabitants, and Biology, archivist of the suffering and the disaster of all that lives.

The current plight of mankind, millions unemployed, the last (and the future) war reveal the *dramatic temperament of Nature*.

Universal Law of Gravitation.- Theory of Relativity. This is well known, it is unnecessary to deal with it here.

[2] Herrera: *El Drama Universal*. "Cronos" and "Emancipación". Mexico, 1932.

Law of Evolution.- "Everything evolves and develops from the homogeneous to the heterogeneous." Progress over the infinity of time is indisputable, and, perhaps, it covers wide horizons, as in the history of Greece, Rome, *etc*.

Universal Law of Life.- There is no such thing as living matter and dead matter, because all alive within the Universe, all that exists can be reduced to the mass or quantity of matter contained by a body, and its life, which is movement; the microcosm or small world, atoms, molecules, is an imitation of the macrocosm or large world and a bond of union and of causation (law by which effects are produced) links the nebulous origin of the sun and the planetary system to the protoplasmic base of the organism, because life encompasses all, from the simplest being to the constellations of the Zodiac, and no sufficient distinction has been established between the living and the inert.

Universal Law of Brotherhood.- All entities or beings in the Universe are derived from electrical fog or Ether and are brothers, but the struggle for life turns them into enemies and dominates mutual aid, which Kropotkin exaggerates, and which is nullified by war, religious hatred, and so on.

CONCRETE LAWS OF PLASMOGENY.

Physicochemical Law.- All phenomena related to an organism, in the past, present and future, had, have or will have known underlying physicochemical forces. There is no vital force, none has been proven, nor is it likely to be so, given the facts of Plasmogeny, by which cells are produced without prior life.

Law of Cytogenesis.- When crystallization comes about slowly in the presence of colloids (glue- or gelatin-like substances), cells and pseudo-beings are formed, which are able to grow and reproduce, albeit in a limited way and, once the technique is mastered, they are likely to multiply indefinitely.

Law of Planetary Life.- When the physicochemical conditions of a planet in the process of cooling down are favourable for the production of solutions of colloids and unstable nitrogenous crystalloids, etc., cells are formed in evolution, with all the transitions between the crystal and the organism, until becoming beings as complex and sensitive as man, perhaps even supermen and unknown others. But it is possible that by increasing sensitivity and suffering, humankind will become extinct by contraceptive means, thereby rectifying an error made by Nature, who errs and modifies her works continuously, without allowing mankind to be regarded as the last and most perfect being, having been unable to attain such - its being incompatible with his treacherous animal atavisms.

Perhaps some unexpected and unsuspected discovery will reform of all mankind, or Plasmogeny will be able to produce artificial, synthetic, fertilized eggs, which give rise to much improved men in an unforeseeable future.

SUPREME IMPORTANCE OF THE NEW SCIENCE.

Should Plasmogeny succeed, Medicine will be able to cure or prevent all diseases, old age and death.

Agriculture, livestock and, in general, animal husbandry, will be replaced by industrial artificial foods and raw materials, as is already the case with dyes, artificial silk, synthetic petrol, rubber, etc. Men will devote their energy to intellectual endeavour instead of debasing themselves, ploughing fields and taming colts or tending to poultry.

Philosophy will extend life and its problems to the Universe, and perhaps, Philosophy and Science will be perfected by man or by artificial brains…

Physical and moral pains will be extinguished forever! Forever!

CHAPTER II

Facts and experiments

PREPARATORY WORK

It would be easy to classify and describe them methodically, but given the unity of Nature, any classification is illusory, and it should be remembered that all matter, even all energy, under the appropriate experimental conditions, gives rise to organic features, of cells, of tissue, of organisms. For example, the electric spark has served Leduc, and others, to represent figures of cell division or mitosis as well as flowery forms of great beauty. Lightning mimics the characteristics of a descending branch.

A thousand whimsical shapes can be found in caves due to calcareous infiltrations, to which people bestow names of people, animals, ghosts, furniture, thrones, organs, etc.

Rocks produce fortuitous profiles of human heads, giants and other objects and beings.

In very cold countries, the ice deposited on windowpanes looks like the fronds of ferns, a variety of flowers and leaves or whimsical lace.

Benvenuto Cellini and other painters and artists have copied the vague figures that appear on old walls or tapestries. One can see countless forms in the clouds, which Hamlet compared to the resourceful character of ingratiating courtiers.

But amid this profusion of appearances, one must choose those that really reproduce living structures; and our enemies mocking criticism is unjust, saying that we see what we want to see, and also unfairly comparing the representations or models of the cell with the phantasmagoria into which feverish delirium converts the forms of shadows and curtains.

The truth is that the problem of Plasmogeny is, simultaneously, morphological, concerning the imitation of forms; chemical, concerning the reproduction of elemental composition; and physical, concerning the reproduction of the physical conditions under which life is produced. We must also address the Geological data and make the experimental methods and laboratory conditions coincide with the natural ones; and in particular with those assumed to exist in the earliest ages of the Earth.

On the other hand, the purpose of this booklet is to clarify concepts and make them available to everyone, and to provide procedures and formulas that can be implemented by readers who want to study the topic: workers, students, enthusiasts and even those of technical profession. I will happily answer any queries and questions, without charge.

In accordance with these considerations, I will discuss the most noteworthy facts and experiments.

METAL PLANTS, PSEUDOPHYTES, OSMOTIC PLANTS OR PLANTS OF TRAUBE AND LEDUC

These are very well known and very easy to prepare, neither laboratory nor microscope are required. One can make these wonders, which have held my attention for many years, with drinking glasses, which can be found in any human habitation, and a few very common reagents that are cheap and plentiful in pharmacies all over the world.

The German Moritz Traube was the first to study them, starting off by preparing osmotic cells. To do this, the gelatin is digested in an autoclave or vulcanizer at 120°C., and mixed with tannin by suspending drops with a rod. They grow like natural cells.

But it is much easier to prepare metal plants. Here is how:

Buy some soluble glass, *i.e.* sodium or potassium silicate, at the pharmacy. Dilute ten parts or grams or any kind of measure, *e.g.*, the contents of a glass, in one hundred of water and stir until completely dissolved. Then place in a glass and drop in some solid bits of iron perchlorite, which are sold in pharmacies. Each fragment is wrapped in a silica membrane of red iron, which soon begins to swell and grow and rapidly rises to the surface of the liquid. It is an interesting effect, because silicate water passes, from outside, through the membrane that develops and forms a thin aqueous solution inside each filament or tube, which is less dense than the external solution at ten per cent. Naturally, the tube is pulled up and rises, while forming a new membrane at the free end. The movement of liquid from the outside to the inside is the phenomenon known as osmosis, taking with it a few molecules of sodium or potassium silicate, formed by the reaction of iron chloride on sodium or potassium silicate. These membranes are said to be semi-permeable, allowing only certain substances to pass through. (Fig. 5)

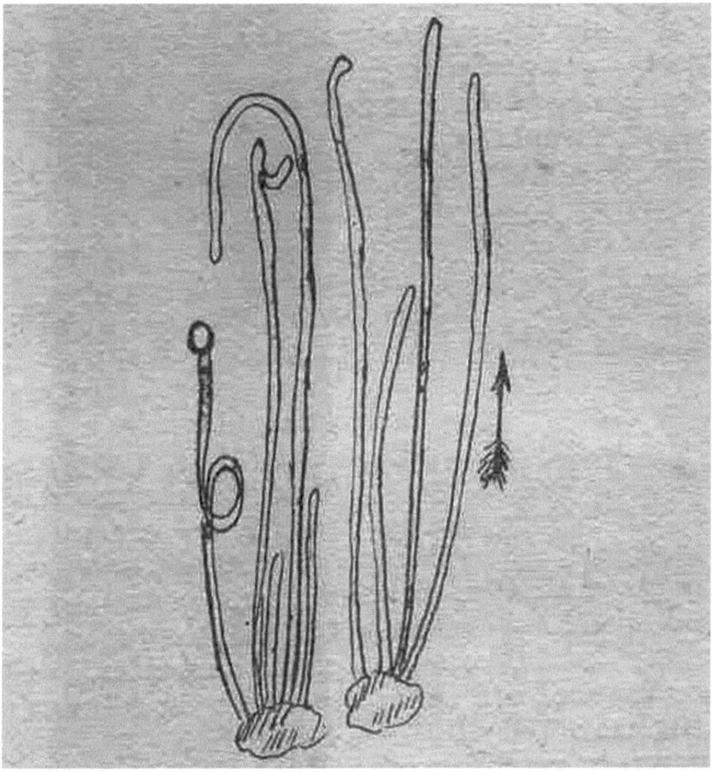

Fig. 5 Particles of iron perchlorite and sodium silicate at ten per cent. Ascending germination.

Modifying the conditions by using artificial seeds of rubber and salts of iron, nickel, cobalt, manganese, etc., and adding water to the upper part of the liquid, or thickening the outer liquid with gelatin, Leduc, Torres, Herrera and others have fashioned a huge variety of metal plants, shell shapes, germinating plants, flowers, worms, etc. (Fig. 6).

They are mainly due to the differences in density between the solution on the outside and that within the membranes, to the bubbles that draw them up and other secondary causes. If one turns the container around, the stalks remain in the vertical position and tangle up or change direction, moving up and down if the container is inverted.

These plants bend towards a bright light when forming. They have been much discussed and furiously attacked by the devout. In fact, they are not really alive and soon become hardened and paralyzed, but demonstrate the role of osmosis and other physical factors in the production of organisms.

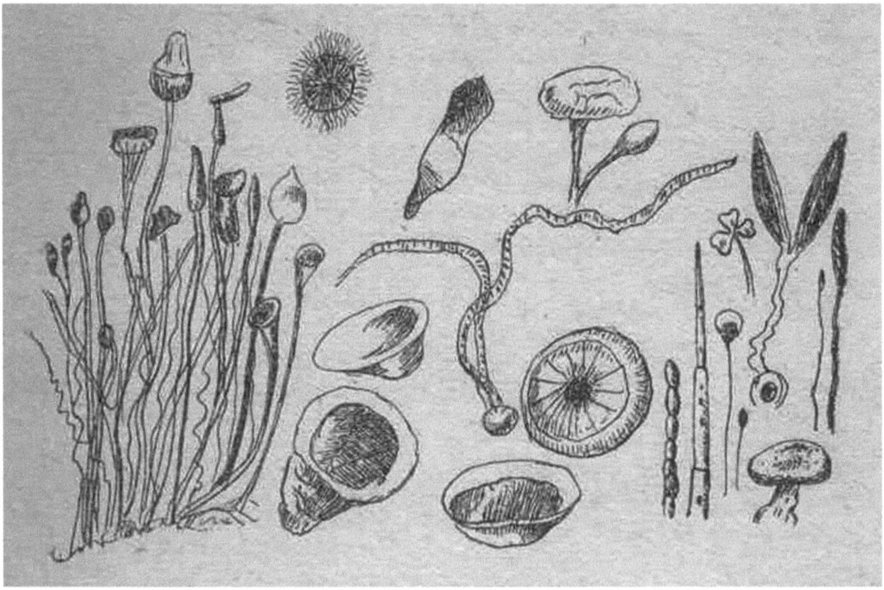

Fig. 6 Traube and Leduc's metal plants, a variety of forms

ORGANIZATION OF SILICATES AND FLUOROSILICATES

Monnier, Vogt and Herrera have produced an infinite variety of cells and organic-like matter by dropping particles of salts, drops of formalin, alcohol, ether or acetic acid into a more or less concentrated solution of soluble silicate or soluble glass. Herrera came to wonder whether silica (sand) might not be the basis of life, as in the gelatinous or colloidal state, combined with calcium chloride and other reagents, it produces a huge range of delicate life-like structures. To do this, fill a pot or Petri dish with colloidal silica and drop some calcium chloride on the edge. The colloidal silica is prepared by neutralizing a highly diluted solution of sodium silicate with hydrochloric acid and dialysing in the machine called a dialyzer, which is sold commercially. It consists of a glass drum, the aperture of which is closed with parchment paper. Immerse in the distilled water of the outer beaker, which should be changed until it no longer gives a precipitate with silver nitrate. In the drum a somewhat murky liquid remains, which is the colloidal silica. It does not allow total crystallization of the calcium carbonate formed as, instead of the particles of calcium chloride, there are others of sodium carbonate, and by double decomposition calcium carbonate and sodium chloride are produced. Microscopic cells and other forms appear. Silica can be found throughout Nature, both inside and outside organisms, and is necessary for life in very small amounts (Fig. 9b).

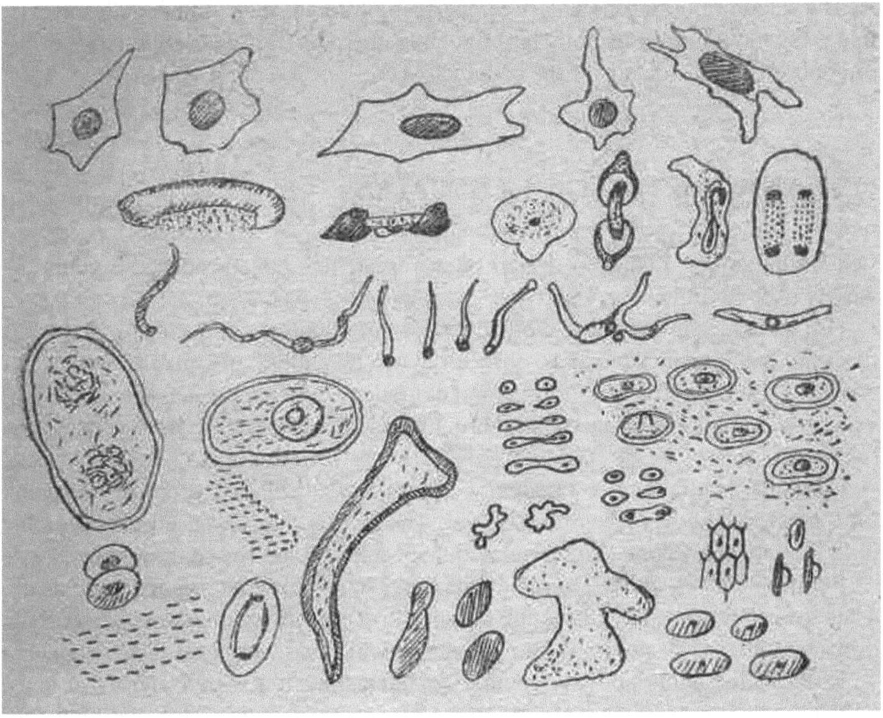

Fig. 7 Colloidal silica, fluorosilicates. Cells, germinations

For years, Herrera has managed to obtain even more perfect imitations of inferior beings, tissues and even the most intimate details of cell division and its essential elements or chromosomes by replacing the sodium carbonate with potassium bifluoride (see Fig. 2, chromatin) (Fig. 7).

This study is ongoing; it is also interesting because silica is very similar to carbon and possibly artificial fluorosilicic beings will be produced in the future, which will be as active and unstable as those that are carbon-based[3].

The best results are achieved with potassium silicate, rather than colloidal silica, with a density of 1100, taken with a hydrometer, 25 cubic centimetres 3 tenths.

Potassium fluoride from Merck, Germany, zero grams 700 milligrams. Formalin, 13 cc. These are mixed and dissolved, and placed in a Petri dish, or preferably between slide and coverslip compressed with a 5-kilo weight. Drop some particles of pure anhydrous calcium chloride onto one edge of the coverslip, at a temperature of 90°C in the oven. After several hours, the calcium chloride will have dissolved in the water of the mixture and slowly seeped in between the slide and the coverslip, producing silicate and calcium fluoride, the former is gelatinous or colloidal, and the latter is crystalloid.

---

[3] See my publication *Una Ciencia Nueva. La Plasmogenia*. Maucci. 1926. Barcelona

The calcium fluoride crystals are imperfectly formed, each of their cells or invisible webs turns into an osmotic bag and altogether affects the characteristics of the protoplasm and the details of the cell nuclei[4](Fig. 7).

ORGANIC DIFFUSION LAB NOTES

Leduc, the author of this work and other researchers have produced a countless variety of these figures.

Diffusion means that a liquid or solid penetrates another or spreads by itself. It is the set of phenomena that can be observed when two liquids are mixed together, for instance, when a saline solution is mixed, gradually, into a greater amount of solvent, or when liquids or gases pass through narrow orifices, or porous membranes (osmosis).

I provide the drawings by Leduc, who recommends placing drops of China ink on a glass slide covered with moist gel. The best is an unexposed photographic plate, which is developed and fixed with hyposulphite and rinsed thoroughly.

Put it on a level, and place a sheet of paper with geometric figures underneath. Then using a pear-shaped China-ink dropper, drip onto the symmetrical points some drops of coloured suspensions, or reagents, in which case use a different dropper.

Leave it for hours and days, and very interesting forms will appear of cells, nuclei, tissues, lower plants, spiders, crabs, Liesegang phenomenon or periodic precipitates formed by parallel lines or bands, which have given rise to copious publications worldwide (Fig. 8a). If the two drops fall together, sometimes, the figure of mitosis or indirect cell division appears, which was thought to pertain exclusively to vital forces; and which is more specifically produced by incomplete crystallization and other means. I shall show this later as although Leduc's figures are admirable, they difficult to obtain without the necessary practice. Thus has the vitalist dogma of karyokinesis or mitosis been overturned (Fig. 8a).

With drops of carbonate and phosphate, of sodium and calcium chloride, one can mimic nacre and pearls, if the diffusion in the gelatin is very slow. Gelatin at 5%, calcium nitrate solution. It is spread and dried, and drops of sodium carbonate and phosphate are added at 5% (Dr. E. Torres).

With coloured liquids, silver, gold, purple of Cassius, potassium ferricyanide and iron perchloride, dyes, etc.., multicoloured mosaics are produced, ornaments of industrial application, on glass, on mirrors or on wood.

Diffusion plays a very important role in the vital processes, associated with incomplete crystallization and other factors.

This can be done in test tubes, and overlapping rings or spirals are produced in gelatin (Fig. 8b). The most common is to make the potassium chromate act on the silver nitrate. Light is involved.

[4] Herrera. *Una Ciencia Nueva. La Plasmogenia.* Maucci. Barcelona. 1926, p. 332 and the next.

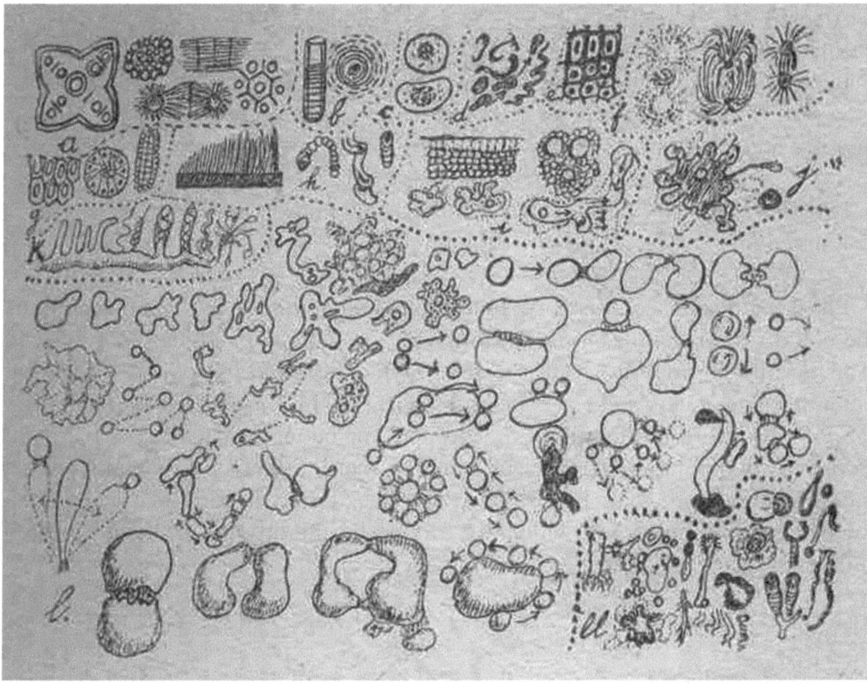

Fig. 8 Different images of artificial life. Colpoides.

UTILIZING COLLODION

Buscalioni, the author of this book and others, use collodion a great deal, which is a solution of guncotton or pyroxylin in alcohol and ether, and must be purchased at a pharmacy, preferably the normal rather than the elastic collodion. Once stained, drip some drops on thick oiled paper, this gives cells with complex colourful nuclei and chromosomes (Fig. 8c).

Otherwise pure alcohol can be added and poured over water: movements, worms, large infusoria that slide rapidly, driven by the alcohol spreading through the water, gyrating; looking like great worms in motion (Fig. 8d). To dye the collodion, add some drops of a concentrated solution of aniline in absolute alcohol.

Buscalioni mixes collodion salts, rubber, chromic acid and other ingredients; applying a thin layer to a slide or glass plate, he blows on it and under the microscope observes: cells, tissues formed by wonderful hexagons, dividing nuclei. These harden (Fig. 8e). It is due to cellular vortices forming liquid axes or nuclei, which are mutually compressed and thus acquire hexagonal forms. The same happens in other simulations and incomplete crystallizations. The author of this book employs a formula that is very easy to prepare:

Alcohol	800 grams
Ether	680 "
Pyroxylin or guncotton	1 to 2 "

Dissolve and add:

Ultramarine blue or ivory black	20 grams
Linseed oil	11 "

Place two drops, equidistantly onto a glass slide, and magnificent radiations and cell division spindles [will appear] (Fig. 8f). To obtain a blue color, instead of using ultramarine blue, starch can be added followed by iodine. By adding water droplets, nuclei and chromosomes appear as the collodion coagulates (Fig. 8f). On absorbing the liquid with a dropper, or better still with an Alvergniat dropper at two points opposite each other on the liquid layer on the glass slide, magnificent mirrored hair-like forms are produced, like the mitosis of cells, with nucleus and chromosomes, which are formed by the droplets of water, located in the middle.

Another formula:

Alcohol at 90 degrees	250 c.c.
Sulphuric ether	240 c.c.
Glycerine	8,400 grams
Ultramarine blue	6 "
Ivory black	1 "
Fine powdered aluminium	5,520 "
Fine powdered gold leaf	5 "
Pyroxylin	6 "

When the ether and alcohol evaporate, cellular vortices are formed, which upon colliding produce the hexagonal shapes in motion, which are tissue-like and can be seen without a microscope. If you cover the dish, they disappear. It is a striking, but very simple, experiment (Fig. 8g). This can also be observed in semi-darkness, by boiling particles of white phosphorus in linseed oil, in a container placed on an electric grill[5]. When the collodion dries, silver and gold hexagons remain. The cells measure up to four centimetres (Fig. 8g). On placing a porcelain disk in the centre, ova and embryo-like forms appear, especially if filaments or ellipses are placed, within which or along whose edges cells gather (Fig. 8g).

[5] Very dangerous. Take precautions.

GERMINATING ALUMINUM

This is one of the better known experiments and has been described in several works. It involves cleaning (delaminating) an aluminium plate, for example, a metal plate purchased in a shop, with caustic soda and a cork until the surface becomes shiny. Then deposit a few drops of mercury or quicksilver onto the metal and rub the delaminated surface hard, with the same piece of cork.

After a few minutes, filaments can be seen, which grow before the observer's eyes and form a kind of very low-density and interesting felt-like material, up to 5 or 6 centimetres long, with the strands ending in points. They grow better in a very humid atmosphere. It is an oxidative or catalytic phenomenon, caused by air in the presence of metals, with the release of hydrogen. According to Wislicenus they grow like natural fibres, forming aluminium oxide. Hydrogen inflates the layers of aluminium oxide or alumina until they fall and are renewed, so that the gaseous pressure gives rise to the germinations[6]. The mercury acts as a catalyst or fermenting agent, thus reproducing the phenomenon that is essential to cell life: fermentation or catalysis. I recommend my readers try this experiment, which is very easy, cheap and highly instructive. The [metal] plate should be washed thoroughly, first with soda, then with a jet of water and afterwards with alcohol, before depositing the drops of mercury. The germinations grow better under a damp glass (Fig. 8h).

COLLOIDS. BUTSCHLI'S OILY AMOEBAE

Around 1885, two eminent Germans, Quincke and Bütschli, the former a physicist and the latter a biologist, a specialist in protozoa (infusoria and other simple microscopic beings) studied the imitation of protoplasm. They managed to obtain extremely good imitations of amoebae (Fig. 1), which continued moving for up to 40 days, in an oven at 40°C, and were composed of very thick rancid oil and potassium carbonate, or salt or sugar. The particles of alkali, salt or sugar, or solution thereof wrapped in a film of forming soap, attract water from without and bring about movements and deformations, due to osmotic flow, as in living beings, where they are accompanied by other phenomena. The simplest way to prepare these amoebae is to grind the listed materials, salt or sugar, with oil and then add water, observing under the microscope. The doughy mixture fills with gaps, vacuoles - it stretches and shrinks - it pulsates, sends out arms or pseudopodia and displays internal granulated currents, which continues for a long time (Fig. 8i).

Bütschli proposed the alveolar theory of protoplasm, assuming it to be composed of alveoli, like a honeycomb, acting like tiny osmometers, a theory that has been refuted and replaced by Chemistry, which assigns to protoplasm a highly complex organic structure, which is invisible.

[6] Alexander. *Colloid Chemistry*. T. I., p. 629. Fibrous Alumina, by Prof. Dr. H. Wislicenus.

But there is much of Bütschli's theory to be maintained, and I think in essence it is correct, because the latest experiments show that the webbing or alveoli of crystals are essential to produce organic forms and their movements. (See the last section on Sulphobes and Cyanosulphobes.)

Bütschli perfected his experiments using the same substances to imitate cells and nuclei, and published several important works. He reproduced the appearance of cell division using air bubbles and gelatine, around which spindles and stars are formed.

I, myself, modified his formulae, making it simpler, as it is difficult to obtain rancid oil with the appropriate chemical composition. To do so, follow these instructions:

Common linseed oil..................... 20 cc
Common hydrochloric acid............ 20 "
Sulphur chloride...........................1 "
Alkaline water............ However much you need

Then sulphur chloride is added to the oil and stirred, with caution, until hot. This takes several minutes. It must be beaten vigorously with a glass rod, because otherwise the oil solidifies in places. Sulphur chloride is used to manufacture artificial rubber. Once the oil has thickened enough, add the hydrochloric acid, with continuous vigorous stirring, and then the chloroform, which keeps the mixture below water, in a suitable vessel, [like] a large crystallizer. Alkalinize the water with ammonia. Upon reaching the bottom, the droplets, injected with pipette, writhe around. Their movements can be seen with the naked eye, undergoing a thousand unique developments, sending out arms, pseudopodia, pulsating, etc. This is a consequence of the osmotic flow caused by combining the inner hydrochloric acid with the outer alkali dissolved in the water, which form ammonium chloride. If a piece of caustic potash is placed nearby, the amoebae will send out an arm to catch it, just like the phagocytes or white blood cells in our blood do with microbes (Fig. 8j).

Oleic acid, used in medicine and sold in pharmacies, produces a variety of forms: cylinders, strands, globules, amoebae, nuclei, nerve cell-like forms, etc. To see this, deposit a drop on a slide under a coverslip and pour drops of alkali onto the edges, forming ammonium oleate, which swells in the alkali water and germinates, twists, looks like writhing worms, amoebae, etc. In this case, liquid crystals are produced, with oily walls, across which the water passes from the outside. Eventually they dissolve. If you use a large amount of oleic acid, dissolved in chloroform or carbon disulphide, and keep it under water, upon adding ammonia you will observe turbulent movements of huge amoebae, for the same reason. I recommend this simple experiment (Fig. k8).

Lehmann and other authors have studied many other examples or cases of organic forms produced by liquid crystals, which I will deal with later.

The experiment with "Colpoids" is linked to this set of assays, and is the most important one before starting those with "Sulphobes".

On observing that the organic forms produced by the infiltration of calcium chloride dissolved in another solution of potassium bifluoride and sodium silicate, retain the appearance of little liquid veins on hardening, I envisaged the technique to produce these changes in the so-called surface tension, *i.e.*, the pressure molecules exert on each other in a liquid. I shall explain this point in greater depth.

When water is sprinkled onto a dusty floor, numerous liquid beads are formed. These are due to the surface tension, as the dust prevents contact between the floor and the water, consequently the latter breaks up into drops due to the attraction between molecules, creating a force or tension on the surface of each drop, as if they were wrapped in a sheet of paper. If spun on a plate or dish the same thing happens, with mercury or quicksilver, which always form droplets. However, when this surface tension is overcome by internal currents or decreases to a point, an arm or pseudopod is sent out from within (Fig. 8j). The following experiment is easy to do: inject quicksilver into quicksilver using a rubber pipette bulb and see how the liquid metal, on penetration, pushes the mercury, which was motionless, and sends forth one or more arms or tubes. Such changes give rise to the movements of the simplest beings or amoebae (Fig. 1), but are more complex due to changes in viscosity, *etc.*

I began, therefore, the experiments by placing glossy white paper under a layer of petrol on a plate and injecting water tinted with ink. A black drop remained on the bottom, moving freely, as water and petrol did not mix and, and petrol is less dense, it stayed on top. Then, with a pin, I drew the water to the edge of the plate and it formed black filaments, like those of a Radiolarian, a microscopic animal having the shape of the sun or a wheel. But this procedure was highly artificial and caused a chemical reaction that disturbed the surface tension. Thus, mixing caustic soda with water and hydrochloric acid with petrol, both causing currents to occur on either side, forming sodium chloride, like in the aforementioned amoebae of oil thickened with sulphur chloride, certain movements and deformations took place. However, hardly any hydrochloric acid dissolves in petrol, therefore I replaced it with acetic acid, achieving stronger- though still insignificant movements. Finally, I tried fatty acids, which on bonding to glycerin form oil, to produce oily soap, the membranes of which enveloped drops of water and soda and thus very strong two-way osmotic flows were produced when the dissolved oil penetrated the petrol, thus becoming more fluid and mixing with the soda, while at the same time excess internal petrol was expelled along with some water. The most commonly used formula is as follows:

French oil, *F. Bétus et Fils, Bordeaux*, fresh, neither rancid nor thick, 50 cubic centimetres, measured with a graduated measuring glass.

Petrol… 100 cubic centimetres. Dissolved by stirring.

Also:

Caustic soda in drums, 12 grams.

Water, heated to about 70 degrees, 100 cubic centimetres.

Dissolve the soda completely in the water. Allow this to cool and then, to dye the soap, add one gram of rhodamine, use a good quality so that lumps do not form. The stained bleach turns very dark purple. Pour the petrol into a porcelain dish and add drops of bleach. Each one divides actively and emits drops, amoebae, worms, infusoria, organic forms that can be seen with the naked eye, and even better under a magnifying lens. They look like living beings moving vigorously, undergoing fragmentation or segmentation, dividing into other smaller ones, this turmoil remains still when examined under the microscope. Do the same in a Petri dish, a circular dish with a lid, and observe under a microscope. If the oil is fresh and not thick or stale, in which case it should be changed, Colpoids appear, imperfect artificial beings, which look like the most common infusoria, called Colpodids (Fig. 8l).

I have published over forty works, in several countries, in which I have cited the publications of various authors, friends and correspondents. The Colpoids were exhibited at the Petroleum Exhibition, in Tulsa, Oklahoma, USA, in October 1930, where there were observed under the microscope by 100,000 visitors. Mr. C.W. Weiant gave a lecture at the New York Museum, with film footage, showing the Colpoids in action.

I will now go on to describe them to some extent:[7]

The preparation sprouts globules in all directions, which arise from the segmentation of the droplets, or amoeboid-like forms differing greatly in shape and size, which are thrown onto one another, as if possessed of a frenzy beyond description, and on touching the contact surfaces bring to bear a mutual, vigorous and violent suction. One can see that something takes place between them, which I first took to be very small electrical sparks. But on observing them under the microscope in complete darkness, they do not emit any light. Further observations showed them to be liquid veins crisscrossing from one "vampire" to another, the suction surfaces undulating to form protuberances and hollows. In some cases tube-like projections form, which penetrate the neighbour and that are clearly not sparks. Finally, the fact has been confirmed that large Colpoids frequently devour the small ones. I have seen the following case: a large one sent out two arms, approaching a smaller one until completely engulfing it. This reciprocal suction, similar to certain movements observed in Vorticella and in some parasitic infusoria, comes in a truly curious variety of forms. For example, two Colpoids suck each other feverishly, each one moving around by itself and in unison, distorting and constantly changing places. On finishing this action, they seem to be satiated and their movements calm down, they finally move away from other and come to rest; but soon after they begin to stir, probably because, via endosmosis, they have absorbed new quantities of oil, fluidized by the petrol. Something seems to stir within them, like a desire or a will, or rudimentary sense of touch (tactism), a primitive intelligence, and again they are pitched onto one another, and with unusual fury renew their vampire-like activities until they become immobile once more. I noticed that if a Colpoid is apparently dead, another close by comes and touches it and begins to suck it, managing to arouse it from its deep sleep, as in Bécquer's rhymes, and both give themselves up to mutual blood transfusion. This effect must be because the intruder alters the osmotic pressure of the sleeper, because if we place Colpoids prepared with seven-percent soda, dyed blue, and fourteen-percent soda, dyed red, in the same oily petrol, we observe that the more highly concentrated ones attack the less concentrated ones, and penetrate them.

---

[7] A. L. Herrera. *Vida y Conciencia Artificiales* (*Artificial Life and Consciousness*). La Medicina Argentina. Year V, N. 54. 1927. P. 1-32 offprints. Figures. Also see *Rendiconti Academia Lincei Roma. Memorias*, 1926. 6th series, Volume II, Part VII.

Logically, this is what should happen, since both should reach the same internal pressure.

I therefore believe that this pressure levelling is the fundamental cause of the struggle, in the same way the parasitic plant called mistletoe (*Loranthus*) cannot accommodate an osmotic pressure other than its own. Another strong influence is the constant inner density fluctuation, which is weakened in the soda on combining it with oil, leading to suction from inside to out and vice versa, as two liquids separated by a thin permeable membrane tend to balance their density.

After several minutes, up to seven hours, at 10°C., on surrounding the beaker with ice, all the Colpoids become motionless, their movements hindered within the soapy crust and lumps surrounding them, as well as the viscosity of the external solution, but on moving their dwelling or adding some drops of oily petrol, they start moving frantically once more.

Often they wrestle just in pairs, but it is also very common for three, four or more to come together, forming chains, rosaries, circles, different shapes, flattening against one other, taking on hexagonal forms. Possibly, the combined suction of the Colpoids is transmitted through all their bodies.

As for shapes and sizes, there is no rule, all being possible, with small, medium, large *etc*., all joining together in the melee. Repeatedly, one can observe that the smaller, more agile ones, charge at the larger ones, collide, bounce back like a ball, attack again, and move back and forth around a large Colpoid as if sizing it up, seeking the most vulnerable spot to sink its teeth, just like a mosquito or flea running over our hand, looking for the weakest spot to bite us, or the most delicious and richest blood. This is when, above all, we realize that the Colpoid is not simply an inert globule driven by osmotic flows, drifting this way or that, carried by differences in level or accidental currents. No, it obviously knows what it is doing; it has a specific purpose, just like the amoebae studied by Lillie, which follow a plan to escape. So too the Colpoids escape obstacles, like the one formed by a wedding ring placed at the bottom of the Petri dish: they clear it with a kind of jump, and carry on their endless motion. Or they run circles around their prey, above and below.

Response to light and other agents.- I believe they seek the light and crowd into the illuminated part of the dish, covered with a sheet of black paper with a slit. Notwithstanding, this may be due to colonies of microbes (*Micrococcus*), which are minute and may have accidentally found their way inside.

They flee from harmful media, acetic acid, for instance. A Colpoid moved forward and reached the spot where I had put a drop of this acid. On approaching, just a short distance away, it stopped, as if reflecting, then with a ripple, fled at full speed. I was amazed. At other times, I have seen a small Colpoid chasing another bigger one, both walk quickly, like children playing catch.

Anaesthetics do not put them to sleep, water dissolves them, gum paralyzes them, the very soap they are made of ends up damaging them if the membrane thickens and encysts, as in natural amoebae, within a capsule; however, if this is broken, after a few days they re-emerge, twisting and turning. Alcohol kills them but they are not paralyzed by potassium cyanide, or quinine sulphate or iron. Conversely, potassium bifluoride destroys them at once, as it is acidic and dissociates the soap. They seem to have the sense of taste, because they taste dead Colpoids or bits of wood impregnated with toxic substances then speed off in search of better prey. One cannot help thinking that *they have, therefore*, a rudimentary consciousness or faculty of knowing. It seems as if they kiss, taste, and suck on each other, or they dislike each other and go away indifferently.

In my articles, cited herein, many observations can be found as well as some data on the electrocapillary movements that are probably caused by currents passing through the membrane and becoming electrically charged, as demonstrated by several authors who have studied these issues in other circumstances. This production of electricity is of great importance and has suggested to me a theory of the origin of thought[8]. Electro-osmosis is very important in Biology, and according to Thales of Miletus, electricity is the soul of the Universe. Thought, and the initial signs thereof, in lower beings or Colpoids, is apparently due to numerical electrical relationships. Ideas consist of electrical relationships and structures, of relationships and structures as yet unknown. Radiograms have already shown us how sound and electric waves send out a symphony, they speak, they sing, which would be impossible without electricity. Lillie and others support similar theories to explain the functioning of the nervous system.

We must, finally, ponder two notions. First, it seems difficult to attribute awareness to Colpoids because we are afraid to - but, even though they are not alive, their actions wonderfully resemble voluntary actions and for this fact alone they are extremely important, because we too are automata, and no unique life force or vital force has been demonstrated, which is not even necessary, as these experiments prove. Second, another fundamental consideration is the silence that has enveloped the Colpoids, despite the fact that their formula has been published in several languages and countries and could even be understood by a child, shows how HUMAN SCIENCES ARE SILENCED AND OPPRESSED BY THE CATHOLIC AND PROTESTANT CLERGY. Their influence spreads far and wide to prevent anyone from knowing such facts, for fear of their downfall. My readers have a moral obligation to repeat these experiments and popularize them, in schools and everywhere.

BEILSTEIN'S AMOEBA AND MODIFICATIONS

Beilstein imitated amoeboid distortion and movements by a procedure that the author of this book had already begun, depositing mercury or quicksilver on a dish and adding a mixture of nitric and chromic acids, which can be done outside the laboratory and without any technical preparation.

The following solution is preferable:

Water............... 500 cubic centimetres

[8] *Estudios*, Valencia. Year X, Number 106, June 1932, pages 23-27.

Potassium chromate......... 20 grams
Nitric acid............... 60 centigrams
Iodide solution
Potassium at 20%............... 17 centigrams

You can have the formula prepared in a pharmacy.

Pour into a dish and add some drops of mercury.

Adding to the solution a pinch of kitchen salt produces a myriad of articulated worms, visible to the naked eye. With a microscope they look like hideous monsters, which writhe, throb, come and go, stay still, then move again, coil and uncoil; fleeing from certain media they are drawn to others and run, leaving a trail with parallel striations. They evolve over four or five hours (Fig. 8ll). Suddenly, they appear to be in possession of a trunk, which sways from right to left, as if searching for prey. Here and there nuclei discs are formed, which expand and contract alternately, due to oxidation, as in natural cells. And most remarkable of all is that this whole microcosm is instantly frozen with ether and other anaesthetics, or with potassium bromide.

This happens due to the oxidation of mercury, formation of nitrates and chromates, mercury iodides and chlorides, which, with the siliceous impurities of the reagents form membranes around each metallic cellule, forming diffusion currents. The membrane ruptures as a result of increased osmotic pressure and changes in surface tension as well as strong internal flows, as demonstrated by injecting mercury into mercury using a pipette, or dropper.

To produce large amoebae, wet some blotting paper, placed on a thick layer of sand or iron sulphide, with the solution. Then, put a large absorbent cotton swab impregnated with the same solution near a large drop of mercury, and the artificial amoeba is drawn up slowly, envelops it with its lateral appendages and seems to engulf it in its own substance, as do natural amoebae and the white blood cells in our blood[9].

Mercurisomes and Hydrosomes[10].- Any dissemination or spreading of liquid over the surface of mercury produces amoeba-like forms. Varying how the reagents come into contact gives rise to a host of biological, electrical, events: wrapping the mercury in little bags made from animal intestine; placing acidulated drops of water onto the surface of the metal or thick rubber onto the mercury; liquid gel; iron sulphide moistened with both chromic and oxide acids; letting the mercury drop from high above the solution or placing it between two glass slides; purifying the mercury to increase its mobility and fluidity or pouring acid from above; cotton impregnated with the acids, especially chromic acid in a highly concentrated solution; covering the liquid metal in collodion. A true museum or microcosm of forms and functional activities arises, such as those in the image: thin membranes of beautiful iridescent colours, an endless succession of life-like forms I cannot describe in this booklet, and which the reader may reproduce easily should he apply himself

[9] Herrera. *El Universal*, Mexico, July 26, 1914.

[10] Herrera. *Boletín de la Dirección de Estudios Biológicos*, volume I, pp. 211-254, figures.

to these fascinating experiments. The hydrosomes (water bodies) are formed by depositing onto the drops of mercury either water or acidulated solutions or, alternatively, salt particles or clay impregnated with certain reagents or aluminium wires. According to Herschell (1824), they are due to the vortices produced by the water chaffing against the mercury, the surface of which maintains an unchanging shape, and which has numerous dynamic points of tangential velocities, due, according to Lipmann, to lines of equal capillary variation. This outstanding physicist has invented special devices to study electrocapillary phenomena, or those produced in thin films. Thus I formed the capillary films which have been thoroughly studied since, by Deveaux and others.

LIQUID CRYSTALS - MESOMORPHIC STATE

Within this huge set of observations, there are countless that show certain states of matter, ranging from liquid, semi-liquid, solid, crystalline and amorphous, or shapeless, forms. All stages and variations have been recorded by a great many authors.

More than a hundred substances comprise liquid crystals, as studied by Lehmann and others, perhaps they exceed five hundred. The mesomorphic state has been the subject of numerous works by G. Friedel and many others.

The easiest way to go about disseminating this Science, object of the present booklet, is to deposit, as I said before, a small droplet of oleic acid, which is sold in pharmacies, between two glass slides, subsequently adding volatile alkali, so that it penetrates round the edges, thus there appears a host of corpuscles, tubes, filaments, organic forms which develop, germinate and move, as can be seen under the microscope (Fig. 8k). I studied them for many years and Lehmann published a great many works on liquid crystals, which he studied using special devices at suitable temperatures. Today, they constitute a well-known chapter in Science and are to be found in the *Enciclopedia de Espasa.*

Recently, Dr. Crile, from USA, believed he had prepared autosynthesizing cells using brain ash and fatty matter extracted from it with ether. What actually happens is that dissolved oleic acid combines with the alkali from the ashes to produce well-known oleates. They are not defined combinations of albumin or protein and fats, and I studied them and made them known some thirty-two years ago[11].

Liquid crystals mimic the most surprising facts of life: structures, forms, movements, evolving and changing state, black crosses when viewed with polarized light.

As shown in the figure (Fig. 9), the crystals display a cube-like shape when seen under X-ray and are made up of alveoli or cells that are invisible to the naked eye and even under a microscope. If the walls of each alveolus are soft and the content is liquid or semi-liquid, plasma is formed that is like natural protoplasm, an osmotic

---

[11] Herrera. Various publications. Historical corrections regarding auto-synthetic cells by Doctor Crile. *Protoplasma*, Leipzig, 1932, Volume XV, Vol. 3, pp. 361-364.

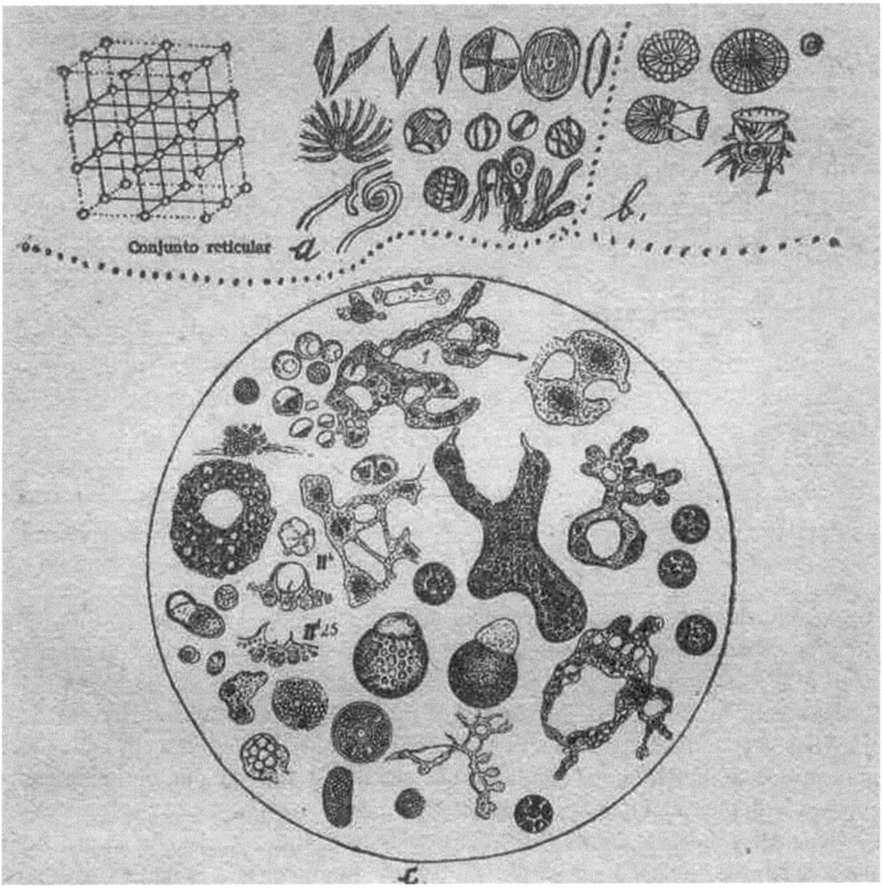

Fig. 9 Crystals, their structure, liquid crystals. Numerous Sulphobes (c)

structure, giving rise to currents and deformations, as in the case of the Colpoids mentioned before, increasing and decreasing the inner pressure of the alveoli and grouping them into rows, filaments and a number of organic forms.

It is likely that this theory is general for all liquid crystals and even the so-called mesomorphic phases, *i.e.*, the intermediate states between the amorphous (formless) and crystalline, which takes two forms: *smectic*, such as soap, and *nematic*, like a filament. These states are always found as follows, according to temperature and dilution:

Crystalline form.
Smectic form.
Nematic form.
Amorphous form.

Thus, between life and death, between crystal and being, there are all these steps, and it is accepted that all beings, including ourselves, are made up of infinitely small crystalline particles (Alexander, Lehman, Della Valle, Herrera, *etc.*).

Incomplete crystallization, hindered or prevented by colloids or gum-like substances, is of paramount importance in Biology and Plasmogeny and there are already countless related works and compiled facts.

Indeed, if salt, for example, tends to crystallize in the presence of gum or egg-white, perfect crystals are not formed, instead globules or stars and numerous organic shapes are formed, due to changes in the mesh or cube-like conformation shown in the figure.

Long ago, Slack and Harting studied this thoroughly. The author of this booklet has also investigated it, showing that the silica gum and egg-white or albumin used, greatly influence the results. Thus, giving rise to a huge number of organic shapes (Fig. 9b).

OTHER WORKS

To prevent this booklet from exceeding its natural length, I shall go on to briefly mention other studies demonstrating the universal tendency of matter to organize itself:

The "cellular tourbillons" or vortices of Bénard Cartaud, Dauzére, Herrera, etc.

When wax and other substances are heated unevenly, hexagons, cells and nuclei appear, due to the vortex or spiralling nucleus that is formed and, on being compressed with their neighbours, create these shapes.

Bénard has been recognised for this research, having received an award from the Academy of Sciences in Paris. The basalt columns inside Fingal's Cave have these hexagonal shapes for the same reason, like the mud on the roads which cracks on drying. Mary and Garrigou proposed a formic acid theory of the origin of life, which is currently the dominant one, and they prepared organic forms - as I will recount later.

Rhumbler has published a special work, within a series of Handbooks, by Abderhalden, concerning a multitude of his own and other authors' studies, especially German studies, reproducing organic shapes and activities, for example, a chloroform drop would engulf a piece of shellac just as our white blood corpuscles engulf microbes.

Dubois, Kuckuck, Burke, H. and others have reported numerous imitations of cells made with radium salts and gelatin, barium salts and gelatin, usually incomplete crystallizations, particularly calcium and barium carbonates in colloidal silica or gelatin. Lecha Marzo, the Spanish histologist, discovered the germination of anilines and alkaloids, which I believe are incidental silicates.

Lillie describes imitations of the nervous phenomena and elements - prepared using objects made of iron, potassium ferrocyanide and calcium chloride.

Jules Felix prepared silicate seedlings, which he presented in an award-winning work at the World Exhibition in Brussels in 1910.

Castellanos, in Havana, publishes a Plasmogeny Newsletter and works on the same science.

In my aforementioned book, Una nueva ciencia: La Plasmogenia (Plasmogeny: A new science), and my annual reviews of the progress made in this doctrine, published in La Semana Médica y La Medicina Argentina (Buenos Aires), I report the wealth of this subject and its continual success and progress worldwide. Dr. J. M. Fontela, of Montevideo, is helping with great enthusiasm to disseminate this new science and its applications to Medicine.

Vibrasomes.- When lycopodium and quicksilver are vibrated on a metal plate, driven from below by a coil that attracts and vibrates them, they produce a great wealth and variety of shapes, forms, emanations, jets, amoebae, geometric figures, simulating lava and flame-throwing volcanoes, rotating spiral nebulae, etc. Lycopodium amoebae climb up steep slopes and fall down the other side, becoming distorted. This is because each lycopodium grain or spore, which is very light, receives an impulse and transmits it to the next one, thus pushing and lifting it from the bottom up, simulating living beings. A series of observations enabled me to understand that these experiments prove Planck's quantum theory of energy, which explains the forces or energies as a kind of equal and invariable pulsation of discrete quantities or quanta. It is therefore very likely that living amoebae also originate from similar vibrations mainly due to chemical reactions[12]. One can simply drop some lycopodium onto the victrola-horn, a kind of cone, of a vibrator and see the most amazing results. The experiment can be done in any workshop where they repair these instruments. The amoebae imitated by vibrating water are very interesting. Everything in Nature is vibrating. Rather than a vital force, instead we find that life comes down to vibration, such as heat and electrical energy, etc.

The vibrators used in massage parlours give very good results. Placing the lycopodium on the disc one can form amoebae, spiral nebulae and other very active forms. Green-stained water between either transparent paper or glass mimics the flow and other aspects of algae, lower plants.

Magnetic spectra.- Iron filings position themselves between the poles of a magnet, creating an image similar to the division of a cell (Fig. 1), and nothing is easier than this experiment. One must sift the filings; meanwhile, the magnet can be made with forceps or tweezers whose tips have been magnetized by rubbing them against a common magnet. One can also replicate the inner parts of the nucleus or chromosome (Fig. 1) with tiny floating magnets.

Future issues.- The ultimate victory will come when we can fully fashion amoebae and other living beings or cells that breathe, feed, possess ferments or diastases, multiply indefinitely, evolve… having the chemical composition and physical properties of natural protoplasm. But we are already on the way to reaching this result and the experiments are progressing very quickly, obtaining highly unstable artificial cells, as discussed in the next section. It is impossible to halt Scienceor

[12] Herrera. *Los Vibroides*. *Institución*, New York, January 1930, page 16, figures.

put a stop to its development, and inevitably biblical dogma will be undermined, held up only by weak branches, on the brink of the abyss and making pessimistic and ridiculous objections - just as it did about aviation, submarines and perpetual motion, which is already almost a fact - produced by radium clocks. What would be unthinkable and absurd is that fanaticism should defeat free Science.

Imitation of brains and other organs.- Threads of plasticine are injected into skulls through a plate with four perforations, representing the spinal cords, and form a cerebral mass with gyri. Otherwise, thin threads can be injected through a rubber bag, or through wire mesh or canvas to form stems, amoebae, coiled-up glands, etc.

The internal pressure of evolving organisms or seeds is similar to this artificial pressure and the plasticine paste represents protoplasm. A model of the thinking organ can be produced thus, and the outraged Faithful have cried out to the heavens, without receiving a response[13].

Injection of gases into glycerated gelatin.- Magnificent solid cells, such as soap suds, plant tissues[14]. The walls of bubbles are compressed and squeezed to form hexagons. Within glassy spheres Radiolarians, those microscopic lace-like beings, are thus mimicked.

Drops of alcohol on silicate with ultramarine blue[15].- Interesting hexagonal cells clumped together, with huge nuclei and chromosomes. Smear some sodium silicate syrups, stained with ultramarine, onto a thick piece of cardboard and then drip some drops of 85° alcohol. This is an easy and colourful experiment. The nuclei have chromosomes, which can be dyed with bone charcoal.

Inorganic cells.- Sodium carbonate, 2 grams; potassium carbonate, 1 gram; colloid silica or highly diluted silicate, 10 grams. This should be left to evaporate in thin layers. Aqueous, deliquescent, nucleated cells, which are dyed and preserved in Canada balsam.

CURRENT WORK

After its almost secular beginnings, Plasmogeny concludes as a transcendental work, concerned with the synthesis or artificial production of protoplasm, in its form, motion and chemical composition.

Over and over again you will find descriptions in the remarkable booklet on biology by Prof. Luis Huerta: La Vida (Biología), CUADERNOS DE CULTURA, especially pages 28-36, which should be read by those interested in this chapter.

Protoplasm, according to well-accepted ideas posed by famous chemists, such as Marcelino and Daniel Berthelot, Bayer, Baly, Baudisch, Moore and Webster, most probably owes its existence to the chemical action of the sun's ultraviolet rays, which reduces carbon dioxide and produces formaldehyde, giving rise to more complex matter like albumin or proteins, by combinations with water, hydrogen

---

[13] Herrera. *Una Nueva Ciencia*, pp. 244-248.

[14] Herrera. *Una Nueva Ciencia*, page 263.

[15] Herrera. *Una Nueva Ciencia*, page 276.

cyanide, otherwise known as Prussic acid, and other materials, as reported in special publications.

For many years, I shared with Raspail and other authors the theory of the origin of life based on photosynthesis, or production through light. Further, when Daniel Berthelot and Gaudechon demonstrated that ultra violet rays of the mercury lamp can replicate the photosynthetic action of plants, I began a series of experiments, as yet unfinished, to produce protoplasm with the substances involved in this mechanism, starting with commercially available formaldehyde or formalin (solution in water).

This research is based, according to Mr. Maynard Shipley, an important U.S. evolutionist, on over sixty years of study by world-renowned scholars[16].

I started off by leaving the formalin in a glass vessel to crystallize imperfectly, giving rise to cell-like spheres, as reflected in communications to the Accademia dei Lincei, Rome.

Afterwards, I undertook some methodical assays, to see how various reagents acted upon formalin, and with ammonium sulphide vapours I achieved an endless array of delicate structures and even moving amoebae and phagocytes or corpuscles that engulf microbes and could, perhaps, be injected into the sick in certain cases of microbial infection. This potential has yet to be investigated (Fig. 9c).

Adding various substances, especially glycerin and nitric acid vapours, the cells are significantly refined when the formaldehyde oxidizes in the presence of the glycerin, giving rise to hydrogen cyanide.

Then I tested how the formalin acted on sulphocyanate or ammonium rhodanide, bearing in mind that some chemists have synthesized amino acids, the base of proteins, with ammonium cyanide and formalin and other reagents.

Sulphur, a component of sulphocyanate, is highly abundant in Nature and in living beings - and it seems indispensable in the production of organic forms - crystallizing imperfectly.

Thus I have come closer to the ideal, steadily closing the gap.

Cells, nuclei, division or multiplication, insightful detail, structures that were believed to be consequence of a vital force, have taken place in my laboratory as I report in this booklet, which I shall send free of charge to whomsoever should request it. The figures sketch these results, which anyone can reproduce using the following formula - one of the many I have studied:

Merck formaldehyde, fresh................................ 7 c.c.
Ammonium thiocyanate fresh, pure and dry........... 3 grams

Boil for ten seconds in a vial or test tube and pour, boiling, onto glass slides or Petri dishes, and observe under the microscope twelve hours later.

It is recommended to do 10 or 20 similar assays and select the best results, since these reagents are very unstable and the smallest circumstance can

[16] Maynard Shipley. *Key to Evolution*. Haldeman-Julius Publication. - Girard, Kansas, U.S. I highly recommend this great work, of reasonable cost.

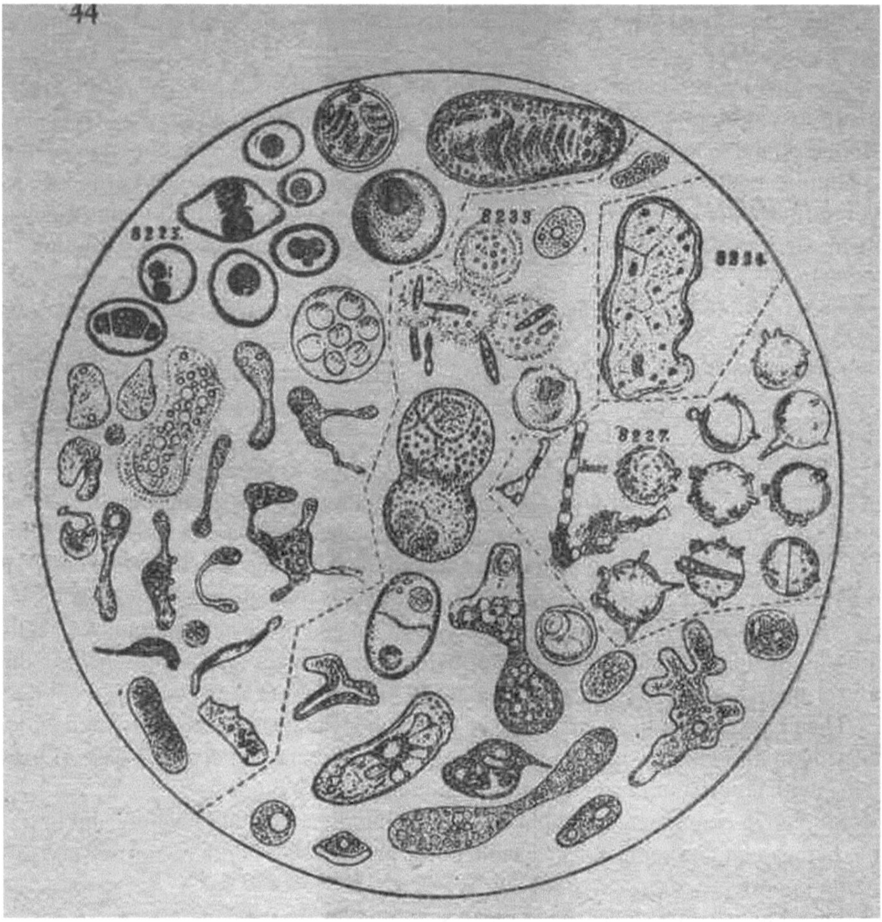

Fig. 10 Cells stained green, blue, yellow and red, remnants of different iron salts. Nuclei in division. Amoebae in anamorphosis and movement. Spores (seeds of lower plants). Flagellates, like little spiny boxes. They have been prepared using Merck formalin and ammonium thiocyanate, boiling, at 82 degrees centigrade, spread in a thin layer on a hot glass plate, measuring 24 by 24 centimetres, slightly tilted, so that the liquid accumulates on one side. In some cases the nuclei have internal filaments or spiremes and reddish chromosomes, like the natural nuclei and cells shown in Fig. 1 Latest experiments. October 1932. Very complex chemical composition, based on carbon, nitrogen, hydrogen and sulphur.

change the phenomenon. In some cases there appear the forms of stars, orbs, chromosomes, and spiremes or filaments, as well as others of nuclei in division (Figs. 9 and 10).

The properties and chemical formula of this product of condensation, studied by the German Prof. Schmerda, are similar to the characteristics chemists assign to polypeptides, the base of albumins.

Results have yet to be achieved with these, or their components, unless they have already started forming in my mixtures. According to my distinguished correspondent, Mr. C. F. Krafft of Washington, albumins should form spirals or whorls, growing via superimposition, like a spring-shaped cylinder to which identical rings are added[17].

OVERVIEW OF THE BOOKET

Plasmogeny artificially reproduces the facts of life, more perfectly with each passing day, and is fully supported by scientific data. Thus it makes way for future discoveries that will bring total freedom, of thought and to Mankind.

[17] C. F. Krafft. *Can Science Explain Life? The Science Press Printing Company*, Lancaster, Pa., U.S. Important booklet that I recommend to my readers. Wöhler (1800-1882), Emilio Fischer, Kossel, Abderhalden and a thousand others have artificially produced urea, polypeptides and many other substances that exist in protoplasm, and which were believed to be the result of a life force emanating from an alchemist God. What ludicrous twaddle! Nowadays, synthetic chemistry surpasses Nature.

MIX
Papier aus verantwortungsvollen Quellen
Paper from responsible sources
FSC® C105338

If you have any concerns about our products,
you can contact us on
ProductSafety@springernature.com

In case Publisher is established outside the EU,
the EU authorized representative is:
Springer Nature Customer Service Center GmbH
Europaplatz 3, 69115 Heidelberg, Germany

Printed by Libri Plureos GmbH
in Hamburg, Germany